Disaster Planning,

Structural Assessment,

Demolition and Recycling

RILEM REPORTS

RILEM Reports are prepared by international technical committees set up by RILEM, The International Union of Testing and Research Laboratories for Materials and Structures. Further information about RILEM is given at the back of the book.

Disaster Planning, Structural Assessment, Demolition and Recycling

Report of Task Force 2 of RILEM Technical Committee 121-DRG, Guidelines for Demolition and Reuse of Concrete and Masonry.

RILEM
(The International Union of Testing and Research Laboratories for Materials and Structures)

Edited by

Carlo De Pauw
Belgian Building Research Institute, Brussels, Belgium

and

Erik K. Lauritzen
DEMEX Consulting Engineers, Frederiksberg, Denmark

Taylor & Francis
Taylor & Francis Group
LONDON AND NEW YORK

Published by Taylor & Francis
2 Park Square, Milton Park, Abingdon, Oxon, OX14 4RN
711 Third Avenue, New York, NY 10017, USA

Routledge is an imprint of the Taylor & Francis Group, an informa business

First edition 1994

First issued in paperback 2011

© 1994 RILEM

ISBN 978-0-419-19190-2 (hbk)
ISBN 978-0-415-51433-0 (pbk)

A catalogue record for this book is available from the British Library

Library of Congress Catalog Card Number: available

Contents

Contributors

RILEM Technical Committee 121-DRG

Dr Susi Buchner, Gifford and Partners, Carlton House, Ringwood
Road, Woodlands, Southampton SO4 2HT, United Kingdom

Mr C. De Pauw, Belgian Building Research Institute, Violetstraat 21-23,
B-1000 Brussels, Belgium

Mr H. Geenens, Belgian Building Research Institute, Violetstraat 21-23,
B-1000 Brussels, Belgium

Dr Yoshio Kasai, Department of Architecture and Architectural
Engineering, College of Industrial Technology, Nihon University,
2-1 Izumi-cho, 1-chome, Narashino-shi, Chiba-ken, 275 Japan

Mr E. K. Lauritzen, DEMEX Consulting Engineers, Godthaabsvej 104,
DK-2000 Frederiksberg, Denmark

Mr M. B. Petersen, DEMEX Consulting Engineers, Godthaabsvej 104,
DK-2000 Frederiksberg, Denmark

Mr J. Vyncke, Belgian Building Research Institute, Violetstraat 21-23,
B-1000 Brussels, Belgium

Preface

One of the greatest technological challenges of our time is to prevent and relieve damages to cities and to protect society from the causes of natural disasters. Another challenge involves the limitation and utilization of the large amounts of building and industrial waste, which are a result of the development in the modern society. Whether the waste originates from clearing after natural disasters or from human controlled activities, the utilization of the waste by recycling will provide opportunities for saving energy, time and resources.

In the early eighties, a technical committee was established in RILEM (Réunion Internationale des Laboratoires d'Essais et de Recherches sur Matériaux et les Constructions/International Union of Testing and Research Laboratories for Materials and Structures), TC-37-DRC on demolition and recycling of concrete. During the work of this committee, which was concluded by the end of 1988, it was established that the field of demolition and reuse of building materials contains some very interesting aspects with consideration to many problems in connection with rescue operations, site clearance and rehabilitation of urban areas overtaken by disasters.

At the end of 1988 UNESCO and the Secretariat of RILEM discussed the possibilities of cooperation concerning earthquake disaster relief in the light of the earthquake in Armenia.

Based on the work in TC-37-DRC and the experiences of the Belgian members in recycling of building materials after the 1980 earthquake in El Asnam, Algeria, a new technical committee was established in 1989: TC-121-DRG on guidelines for demolition and reuse of concrete and masonry. According to the terms of reference for this committee, the working program comprises:

- The examination of quick and safe removal and demolition of major concrete structures and buildings after structural collapse.

- The preparation of a state-of-the-art report on the guidelines.

- A recommendation for international guides for site clearing after earthquakes etc.

At the first meeting of the committee in Copenhagen, September 1989, it was decided to establish two task forces to cope with the objectives of the technical committee, whereof Task Force 2 was established with following terms of reference:

"The Task Force will prepare a State-of-the-art report on site clearing and demolition of damaged concrete structures with respect to the reuse of concrete and protection of the remaining structures. Special emphasis should be placed on earthquakes and war damaged structures."

Members of the Task Force:

Dr C. De Pauw, General Director, BBRI, Belgium (Chairman)
Dr S. Buchner, Gifford & Partners Ltd, England
Professor Y. Kasai, Nihon University, Japan
E. K. Lauritzen, DEMEX Consulting Engineers, Denmark
J. Vyncke, BBRI, Belgium (Secretary)

Program of the Task Force:

- Collection of literature.
- Summarization of literature.
- Preparation of report.

The collection and summarization of literature took place in 1990 and 1991, and preparation of the report in 1992 to 1993. TC-121-DRG plans to finish its work in the Autumn of 1993 with the third international symposium on demolition and reuse of concrete and masonry, to be held in Denmark, 25th–27th October 1993.

1

INTRODUCTION

Erik K. Lauritzen
DEMEX Consulting Engineers A/S, Denmark

1.1 Background

According to the Brundtland Commission's report "Our Common Future" [1], it is apparent that there is a growing understanding of the necessity to limit the amount of waste which is today being generated in the construction and building sector. After recycling, these wastes may be utilized as substitute for primary raw materials thereby conserving our natural resources.

A large part of waste derives from; demolition, rehabilitation, and new construction following normal development as well as natural and technological disasters. For example, the production of building materials and goods involves ready mixed concrete, concrete elements, articles of wood etc., which can be classified as industrial waste.

At present very limited amounts of building waste are recycled, the majority being deposited or used as fill. Since building waste amounts are constantly increasing, there are many reasons for focusing on methods which will promote an increase in recycling of demolition and building waste (dumping fees in Europe and USA are typically from 20 - 50 US$ per ton). Present results in Europe show very favourable recycling possibilities in this field.

From a purely economical point of view, recycling of building waste is only attractive when the recycled product is competitive with natural resources in relation to cost and quality. Recycled materials will normally be competitive where there is a shortage of both raw materials and suitable deposit sites. With the use of recycled materials, economical savings in transportation of building waste and raw materials can be obtained.

This is especially apparent where there is a local combination of demolition and new construction, making it possible to recycle large amounts of building waste at the work site or in the vicinity.

Application	Project example	Waste material
Aggregate in new concrete	Concrete roads	Crushed concrete
	Runways, taxiways and aprons	"
	Concrete pavements in general	"
	Concrete sewage pipes	"
	Concrete culverts	"
	Bridges	"
	Harbour constructions	"
	Environmental plants:	"
	_ water treatment plant	"
	_ pumping station	"
	_ fertilizer tanks	"
	_ refuse disposal plant	Crushed concrete/brick
	Buildings (residential, commercial):	
	_ foundations	Crushed concrete/brick
	_ floors	"
	_ horizontal divisions	"
	_ walls	"
	Foundations in general	"
Aggregate in new asphalt	Base course materials in pavements and yards	Crushed concrete
Unbound base course	Bicycle lanes	Crushed concrete/brick
	Pavements	"
	Field roads	"
	Forest roads	"
	Internal building site roads	"
	Primary roads	Crushed concrete/brick /asphalt
	Secondary roads	"
	Runways, taxiway and aprons	"
	Parking lots and other yards	"
Fill material	Cable trench	Crushed concrete/brick

Fig. 1.1 Examples of the possibilities for reuse of brick and concrete waste materials.

The need for waste reduction and resource management is indicated in World Bank publications, for example: "Environmental Management in Development" by Michael E. Colby, 1990 [2], and "Investments in Solid Waste Management - Opportunities for Environmental Improvement" by Carl Bartone, Janis Bernstein and Frederick Wright, April 1990 [3].

In the last decade much research and development concerning the reuse of building materials has been carried out. However, feasibility studies for the implementation of recycling tools and the use of recycled materials in urban development and rehabilitation are lacking.

Reuse of building materials has always been common practice, whereas processing and recycling of materials to primary resources (e.g. crushing of building rubble to aggregates and melting scrap metal to construction steel) has only been applied in highly industrial societies.

The research and development in Europe within the field of solid waste management and techniques has proven the applicability of recycling and reuse of waste materials in the building and construction industry. In a few countries, e.g. Denmark and The Netherlands, standards for the use of recycled aggregates for concrete in passive environmental class are recommended.

It is also well documented that aggregates based on bricks or masonry wastes perform very well in most concrete for structures which are not exposed to saturation and frost/thaw conditions.

In a situation of free production costs, the choice between recycled and natural materials depends upon quality. The quality of concrete with recycled aggregates is the same as or a little lower than that of concrete with natural aggregates.

With larger recycling projects, such as urban development, renovation of motorways, or clearing of war/disaster related damages, the economical model as shown below will be dominated by transportation costs. These transportation costs involve the removal of demolition products and supply of new building materials. In these cases the use of recycled materials is very attractive.

In the USA, the Federal Highway Administration often performs crushing and recycling of concrete surfaces. For example, during the renovation of a seven mile highway stretch in Wyoming, 1985, the aggregate was a mixture of recycled and natural materials, thereby giving a total economic saving of 16%. With the use of recycled materials, a saving of 35,000 - 100,000 US$ per mile can be obtained in comparison to traditional methods.

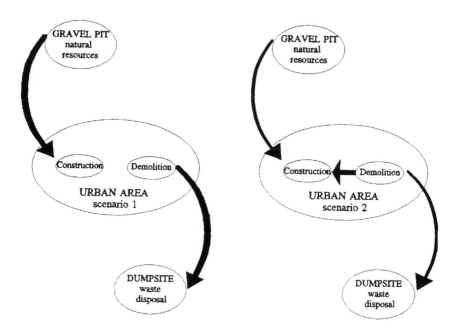

Fig. 1.2 Macro-economical model showing two urban renovation/rebuilding scenarios. In scenario 1, all building waste is removed and deposited and only new raw materials are supplied to the site. Scenario 2 shows how a considerable amount of the demolition waste is reused as raw materials in the new construction. Only refuse waste, including polluted waste, is removed and deposited, thereby reducing the demand for new raw materials. The economical savings of scenario 2 are primarily related to transport costs supplemented by a time profit.

Results from recent projects in Denmark show that a total saving of approximately 15 US$ per ton can be obtained with recycling of materials when compared to demolition and dumping of material by traditional methods.

There is no doubt that the European and American research and development results and experiences can be transferred to other parts of the world to enable natural (primary) raw materials to be replaced by recycled materials, especially in urban renewal and rehabilitation construction projects.

In 1980 the Belgian Government sponsored a pilot project for the recycling of building waste after the Algerian earthquake [4]. From a technical point of view the pilot project was successful, however, it did not continue for several reasons, amongst which political and sociological can be mentioned.

After the earthquake in Armenia 7 December 1988, the German Red Cross sponsored the installation of a recycling plant in Armenia. Referring to available information [5] this recycling project met a lot of technical problems, e.g. bad weather, frost, lack of energy and water etc. However, the project is not yet fully reported and the latest results are as yet unknown.

In late 1988, UNESCO and the Secretariat of RILEM (Réunion Internationale des Laboratoires d'Essai et de recherches sur Matériaux et les Construction) discussed the possibilities of cooperation concerning earthquake disaster relief in the light of the earthquake in Armenia.

In 1989 RILEM set up a technical committee RILEM TC-121-DRG Guidance for Demolition and Reuse of Concrete and Masonry. In order to meet the requirements of UNESCO this committee decided to deal with demolition and recycling after disasters as its main objective.

On the basis of the European and American experiences it may be concluded that the present technical level of know-how in the field of recycling and reuse of building and construction materials is sufficient to implement these techniques into the continuous rehabilitation process which takes place in the urban areas of developed countries.

Therefore, it is recommended to investigate and prove the feasibility of the techniques as integrated parts of Systems for Disaster Relief in the event of disasters and for urban development projects.

1.2 Integrated disaster relief and disaster vulnerability

In accordance with The World Bank's terminology, disaster basically includes all occurrences, natural or technical, resulting in damages to local communities to such an extent that outside help is required, ref. Kreimer & Munasinghe [6].

The disaster concept is normally associated with typical natural disasters, eg. earthquake, flooding, volcanic eruption, drought, famine. In addition there are medical disasters (epidemics) and man-made (more or less self-inflicted) technological disasters, such as armed conflicts, contamination, shipwreck, air crashes, industrial disasters.

According to the mutual insurance company Muncher Ruck, a disaster is classified as such in insurance terms when the loss of life is more than 1,000 or the economic losses are significant [7]. Nobody doubts that the eruption of the dead volcano Pinatubo in 1991 was a disaster although actually only one life was lost directly caused by the eruption, whereas incredibly large arable areas were totally destroyed.

The World Bank places great emphasis on linking environment and population growth/density with disaster risk, which is described in more detail in the World Bank's publication "Environmental Management and Urban Vulnerability" [8]. Because disaster vulnerability is closely related to man-made activities, it has been found - among other facts - that on average over a 10 year period natural disasters have cost the lives of 250,000 people and caused damages worth nearly US$ 40 billion per year. The World Bank furthermore estimates that the frequency of earthquakes, volcanic eruptions etc has increased in recent years due to the Greenhouse effect caused by increased CO_2 emission.

With reference to UN General Assembly Resolution No. 42/169 of 11 December 1987, UN has declared the 10-year period 1990-1999 The International Decade of Natural Disaster Reduction (IDNDR). It is, according to The World Bank, the aim during this 10-year period to improve the individual countries' abilities to deal with the results of natural disasters, to stipulate directions and strategies for using the existing knowledge on the subject, and to develop and research waysof reducing loss of life and property. In addition it is the intention to modify existing and seek new information regarding risk-evaluation, predictions, preventions and preparedness for natural disasters and to develop programmes for technical assistance, technology transfer, demonstration projects, education and training with regard to specific threats.

IDNDR emphasises as an example the contribution regarding reduction of the over-populated cities' vulnerabilities (i.e. Mexico, Sao Paolo, Manila etc.) and points out the demand for analysis of the risks of epidemics, hunger etc. caused by damages to supply lines, infrastructure etc.

Apparently OECD does not give IDNDR much importance: its publication "The State of the Environment 1991", [9] has only a limited reference to natural disasters and states that there is no scientific evidence to assume that the frequency of natural disasters will increase or decrease in the future. According to a German report, IDNDR is supported by the EC and NATO and by several countries, eg. USA, Japan, United Kingdom, France and Germany.

The demand for disaster assistance and relief is global, and this demand is much more comprehensive and long-term than that for emergency humane relief. There is a great need for assistance, especially in the form of preventive and mitigative measures; the demand for mitigative measures by nature being short term. In order to give effective disaster assistance it is necessary to look at the assistance as an integrated entirety - Integrated Disaster Management - based on co-operation across professional and political beliefs.

Referring to Fig. 1.3, the concept of disaster assistance can, as an example, be divided into a 6 phase chain of events, or into 5 phase cycle [10]. Thus there is a very differentiated demand, which depends on the actual character of the disaster and the geographic position. The different phases mentioned are completely differently structured, regarding both financing and management.

The preventive and mitigating arrangements etc. carried out before the disaster are normally done by the local community under supervision and with external financing.

Depending on the type and size of the disaster, the initial relief work will typically be managed by the local community under de-centralized management. This means that the relief is delivered and received subject to the prevailing circumstances and conditions. As these are normally situations of extreme stress, the relief work will often be characterised by organisational and managerial problems, a general shortage of goods and resources, failing communications etc. Delivery of supplies and external assistance to the relief work as well as medical assistance depend mainly on the international relief organisations and their financing.

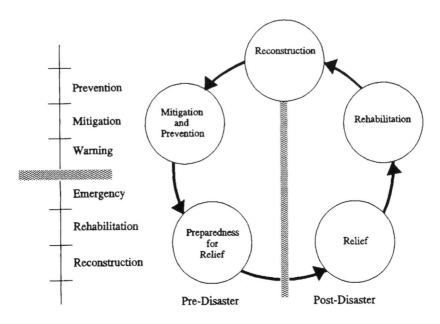

Fig. 1.3 Illustrative examples of phasing in a disaster development, [10]

When entering the next phases regarding temporary arrangements and re-building, more control of the organisation and management as well as financing will develop. Depending on the size of the disaster and the technical and economical capacity of the country in question, the national government, normally in co-operation with development banks, will complete a special reconstruction programme, financed by one or more of the development banks.

In The World Bank's policies regarding disaster assistance emphasis is placed on preventive arrangements and analysis of the 'anatomy' of the disaster and its consequences. It is, as an example, important as soon as possible in the disaster development to aim at a reconstruction of the infrastructure and supply lines. The World Bank has in 1992, commenced a project with the aim of supporting national and local authorities in risk analysis, to develop policies regarding preventive and supportive arrangements, to identify potential projects for World Bank and international financing and to train and develop educational material concerning relief and self-help. Special emphasis is placed on planning and disaster preparedness, including organisational, technical and socio/psychological preparedness, both in the previous mentioned project and in the IDNDR programme.

It should furthermore be pointed out that The World Bank emphasises the demand for prevention and solving the environmental problems connected with a disaster. In addition it should be noted that The World Bank places great importance on the input being aimed at an integrated solution based on "cleaner technology" views.

1.3 Description of the work

The work of Task Force 2 of the RILEM TC-121-DRG has mainly been based on the work made by the Japanese member of the Committee and reports prepared by Japanese researchers and other information collected in Japan.

Additionally, information, experiences and investigations have been collected by the other members of the Task Force 2 of the Committee from participation at conferences and investigations of recent disasters, for instance:

- Earthquake El Asnam, Algeria 1980;
- Earthquake Mexico, 1985;
- Earthquake Armenia, 1988;
- Earthquake San Francisco, 1989;
- Earthquake Philippines, 1990;
- Volcano eruption Philippines, 1991;
- Kuwait-Iraq war, 1990 - 1991;
- Earthquake Turkey, 1992.

An extensive list of literature and other references which have been reviewed with special attention to demolition and recycling of building materials is presented in Appendix I.

Presentation of the most interesting references is shown in literature review Appendix II.

References

[1]   Brundtland Commission Report, "Our Common Future". United Nations, 1987.

[2]   Colby, Michael E, "Environmental management in development countries. World Bank discussion papers 80, 1990.

[3]   Carl Bartone, Janis Bernstein & Frederick Wright, "Investments in Solid Waste Management". Infrastructure and Urban Development Department, The World Bank, Washington, 1990.

[4]   C. De Pauw, "Recyclage des Decombres d'une Ville Sinitree", C.S.T.C. - Revue/N°4/ Dec 1982, Bruxelles, 1982.

[5]   Dipl.-Ing. Steinforth, "Einsatz einer Baustoff-Recycling-Anlage im Erdbebengebiet Armenien/UdSSR". 6. Symposium Recycling-Baustoffe, Cuxhaven, Germany, 1990.

[6]   Alcira Kreimer & Mohan Munasinghe, "Managing Natural Disasters and the Environment", Environment Department, World Bank, Washington, 1990.

[7]   Munich Reinsurance Company, "Earthquake Mexico '85", Münchener Rück, München, 1986.

[8]   World Bank Conference, "Environmental Management and Urban Vulnerability". Environment Department, The World Bank, Washington, 1992.

[9]   Organisation for Economic Co-operation and Development Report, "The State of the Environment". OECD, Paris, 1991.

[10] Frederick Krimgold, "Pre-disaster planning". Department of Architecture, KTH Stockholm, Vol. 7, Sweden, 1974.

2

Disaster Rescue and Restoration Countermeasures in Japan

Yoshio Kasai
College of Industrial Technology of Nihon University

Kunihiko Hirai
Japan Urban Safety Research Institute

2.1 Introduction

"When a disaster, such as war, major earthquake or fire, occurs, how would human lives be rescued from a collapsed building? What would be done about rehabilitation? Explain the systems and specific countermeasures." This was the theme given to me by the committee.

This problem cannot be solved outright. This is because a prescription cannot be written for a world where ethnic, cultural and economic conditions are different. As a resident of an island country in the Far East, I will report on the situation of disaster countermeasures for large-scale earthquakes which we frequently experience. I hope that this will be examined by the committee and contribute to the preparation of the guidelines.

The most important theme in disaster countermeasures is its prevention; major fires can sometimes be prevented by being careful, however just being careful will not prevent major earthquakes from occurring. A disaster prevention countermeasure for earthquakes would be to build an earthquake resistant structure based on a hypothetical earthquake and to implement countermeasures and drills predicting the coming of an earthquake as early as possible.

After the occurrence of an earthquake, a powerful disaster countermeasure organization must be formed quickly. The public sector and private personnel must cooperate and implement emergency countermeasures and grapple with disaster rehabilitation. This report is prepared from such a perspective.

In the following I will outline the sections in this paper respectively.

1. System of disaster relief in the occurrence of a disaster

In 1961 the "Disaster Countermeasures Basic Act" was established. This act became the basis on which the whole of the disaster countermeasure law was based. It regulates the Japanese Central Disaster Prevention Council which discusses basic disaster prevention items, emergency countermeasures in times of disasters, and application criteria for the Disaster Relief Act.

2. Systems for disasters of extreme severity

To deal with disasters of extreme severity, the "Act for Disasters of Extreme Severity" was established in 1962 to make it possible for the national government to give special financial aid. Here, issues such as the designation procedures and criteria for the application measures for disasters of extreme severity and the content of relief will be discussed.

3. Issues on the present disaster relief system

Issues such as the procedural problems which arise when a disaster has occurred and the adjustments between various self-governing bodies when a disaster of extreme severity spanning a wide area occurs are pointed out.

4. Role of the Self Defence Forces

The Japanese Self Defence Forces are considered to be a group of relief activity specialists in the occurrence of disaster. They are responsible for preparing the "Operational Plan for Disaster Prevention" particular to the Self Defence Forces. Furthermore, legal bases for disaster dispatching requests are discussed.

5. Lifesaving and victim relief activities

Important points in themes such as saving and rescuing disaster victims, public burdens for emergency rehabilitation, hygiene sanitation, and maintenance of social order will be discussed.

6. Exercise of public authority for the dismantling and handling of victimized buildings

In removing a victimized building, the use of public authority in the duty of administration on legal grounds and the disposal of privately owned victimized structures is discussed.

7. Rehabilitation of lifelines

The institutions sharing the responsibility for the rehabilitation of lifelines such as water and sewage, gas, electricity, communications, roads and railways are identified. Furthermore, the relationship between the number of days after the outbreak of disaster and the rate of rehabilitation of various lifelines for the past two major earthquakes is illustrated.

8. Cases of disaster relief and restoration activities

Such problems of scale, rubble disposal by the Japanese Self Defence Forces, steps in selecting a rubble disposal site, and garbage designation of privately owned areas concerning the Great Fire of Sakata are raised.

The Miyagi Earthquake was an earthquake which attacked Sendai City, a city with a population of 640,000. As a result, there were 4,250 completely or half destroyed houses. Water, electricity and gas supplies were halted but the citizens dealt with the situation coolly and thus evaded a panic.

9. Study of future Kanto Great Earthquake countermeasures

The study begins with a classification of the resulting damage from the Kanto Great Earthquake in 1923. The two main aspects of damage experienced due to the disaster are illustrated and the ensuing garbage disposal plan for the affected Tokyo area is presented.

10. Dismantling, rubble disposal and recycling of victimized structures

This is an important theme in the RILEM 121-DRG, Task Force II work:

1. The degree of damage of a victimized structure must be classified. A criteria of classification must be established in normal times and competent assessors must be trained. To be able to make the appropriate assessments in the occurrence of a disaster, it is important to be prepared.

2. Subjects such as the development of technology of rescuing survivors from collapsed buildings and the use of dynamite in the rapid dismantling of collapsed buildings are studied.

3. The measures and problems involved in the disposal and recycling of rubble are discussed.

The preparation of this paper is mainly the work of Kunihiko Hirai, Secretary-General of the Urban Safety Research Institute. Furthermore, it is based on the cooperation of the individual members of the Architectural Institute of Japan RILEM subcommittee (chief examiner Yoshio Kasai) workshop (121-DRG). I would like to express my gratitude by hereby citing their efforts.

2.2 System of disaster relief in the occurrence of a disaster

Summary of disaster countermeasures

After World War 2, Japan was stricken by various major disasters. Such experiences resulted in reports and recommendations for reinforcement and consolidation of the national disaster prevention system.
In the meantime, the Ise Wan Typhoon of September 1959 resulted in a major disaster, which enhanced the motivation among those concerned to proceed in developing a nationwide, comprehensive and objective administrative system for disaster prevention. As a consequence, the "Disaster Countermeasures Basic Act" (under the jurisdiction of the National Land Agency) was promulgated in November 1961.
   The Disaster Countermeasures Basic Act defines the basic administrative policies with regard to the following aspects:

1. Clarify the jurisdiction and responsibility for the administration of disaster prevention.
2. Establish the total disaster prevention system encompassing the national and local government.
3. Administer the disaster prevention measures purposely.
4. Strengthen the countermeasures for disaster prevention.
5. Prepare for the expedited and appropriate countermeasures for the emergency relief on the damages inflicted by disaster.
6. Execute an expeditious recovery and restoration from the disaster damages.
7. Assign the fiscal responsibilities to appropriate parties.
8. Execute necessary actions while the disaster is being inflicted.

Central Disaster Prevention Council
With the Prime Minister as chairman, and the heads of the Designated Administrative Organs, and personnel of similar status, as members, National Government has established the Central Disaster Prevention Council to deliberate important matters relating to disaster prevention. This council is making every effort to promote extensive and comprehensive countermeasures against disasters.

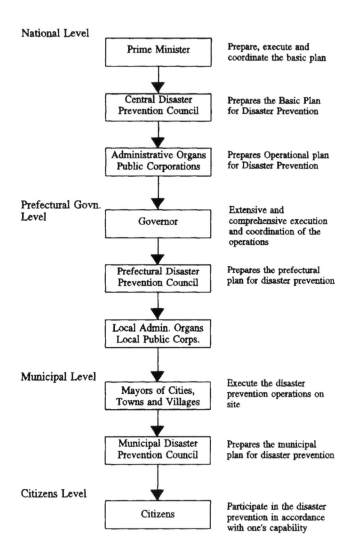

National Level

Prime Minister — Prepare, execute and coordinate the basic plan

Central Disaster Prevention Council — Prepares the Basic Plan for Disaster Prevention

Administrative Organs Public Corporations — Prepares Operational plan for Disaster Prevention

Prefectural Govn. Level

Governor — Extensive and comprehensive execution and coordination of the operations

Prefectural Disaster Prevention Council — Prepares the prefectural plan for disaster prevention

Local Admin. Organs Local Public Corps.

Municipal Level

Mayors of Cities, Towns and Villages — Execute the disaster prevention operations on site

Municipal Disaster Prevention Council — Prepares the municipal plan for disaster prevention

Citizens Level

Citizens — Participate in the disaster prevention in accordance with one's capability

Fig. 2.1 Planning process for Central Disaster Prevention Council.

Basic Plan for Disaster Prevention
The Basic Plan for Disaster Prevention was prepared by the Disaster Prevention Council in 1963, and it has been used as the bases for all future disaster prevention programs that have been developed.

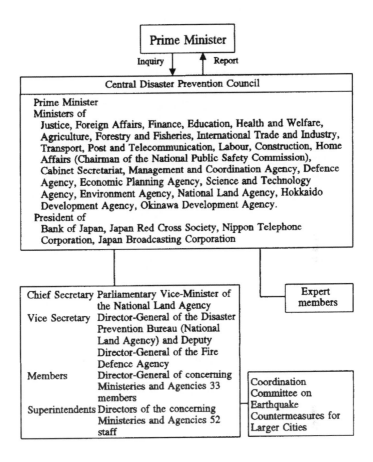

Fig. 2.2 Organization of Central Disaster Prevention Council.

Operational Plan for Disaster Prevention
The Operational Plan for Disaster Prevention is made up of individual bodies in accordance with the Basic Plan for Disaster Prevention mentioned above.

Local Plan for Disaster Prevention
This plan defines in concrete terms the disaster prevention operation which a local disaster prevention organization is required to perform in their respective autonomous areas, whether on a prefectural (regional), city, town or village level. The plan is structured in accordance with local situations and requirements.

Designated Administrative Bodies and Public Corporations
There are, as of April 1992, 29 designated Administrative Bodies in the national organizations for disaster prevention. These consist of The National Land Agency and other ministries and agencies of the national government, as well as 37 Designated Public Corporations such as the Japan Red Cross Society, Nippon Telegraph & Telephone Corporation.

Designated Local Administrative Bodies and Public Corporations
There are some local administrative bodies (local branches to the national government) and local public corporations which are locally designated, for the preparation and execution of disaster prevention programs.

Organization of Disaster Countermeasure Headquarters

Upon the occurrence of a disaster, or when one is anticipated, the national administrative bodies, local government, and public corporations will proceed to take various emergency countermeasures to prevent or control such a disaster.

Should an earthquake occur, the municipality will at first establish a Municipal Headquarters for Disaster Countermeasures to execute emergency operations. If necessary, the prefectural government would establish its Headquarters for Disaster Countermeasures.

In the meantime, The National Land Agency will assess the severity of the disaster, and if deemed necessary will call up the Meeting of Ministers and Agencies related to Disaster Countermeasures for exchanging pertinent information. Should the need be recognized, the national government would establish the Headquarters for Major/ Extraordinary Disaster Countermeasures, in accordance with the Disaster Countermeasures Basic Act, and proceed to execute comprehensive disaster emergency countermeasures.

In order to carry out such emergency countermeasures smoothly and promptly, the administrative bodies at each level (from the national - prefectural - municipal) will have prepared Plans for Emergency Countermeasures in order to cope with disasters.

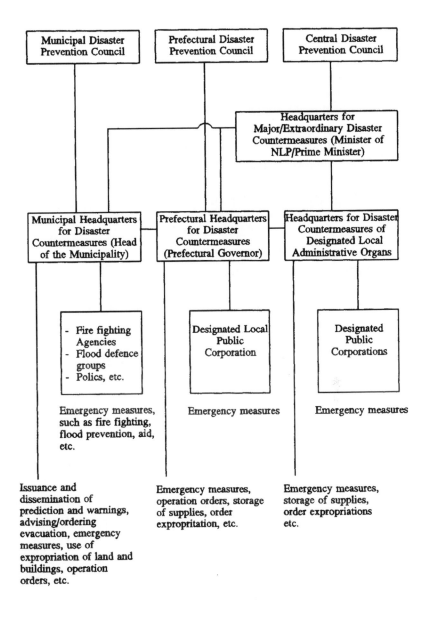

Fig. 2.3 Main Emergency Measures to be taken in the event of a Major Disaster.

A typical case of planned Emergency Measures against Disaster
(Tokyo Metropolitan Plan for Disaster Prevention (Earthquake Division
Version)) is given below:

1. Establishment of the system for emergency activities, (the
    operating of the headquarters, etc.).
2. Collection and dissemination of information.
3. Application of the Disaster Relief Law.
4. Mutual assistance, and the plan for requesting dispatch.
5. Fire-fighting and disposal of dangerous material.
6. Flood defence and measures against tsunami.
7. Security guard and traffic.
8. Emergency evacuation plan.
9. Rescue and ambulance plan.
10. Medical and first-aid plan.
11. Plan to supply drinking water, foodstuff and other essential
    provisions for livelihood.
12. Emergency transportation plan.
13. Plan for cleaning, prevention of epidemics and disposal of human
    remains.
14. Provisional housing plan.
15. Plan for temporary schooling, financing and employment.
16. Provisional measures to rescue and secure life-line facilities.
17. Provisional measures for operating public facilities, etc.

Disaster Relief Act

The Nankai Earthquake, which occurred in 1946 shortly after the
Second World War, caused great damage in a large area covering the
whole of Western Japan. Due to the Victim Fund Support Act at the
time, shortcomings such as varied prefectural unit relief expenses and
frequent lack of coordination between institutions concerned in the
implementation of relief activities and in the procurement of supplies,
became evident. Thus, the Disaster Relief Act was established in 1947
as radical legislation to compensate for these shortcomings.

The Disaster Relief Act established regulations for overall relief
activity to systematically implement emergency and temporary relief for
disaster victims. Furthermore, it clarifies the relative burdens of
expenses between the National Treasury and the prefectures.

1 Application standards for the Disaster Relief Act
(Either A or B below must apply)

A	When more than a fixed number of houses in proportion to the population of a prefecture or municipality has been destroyed (for a prefecture of less than 1 million or a municipality of less than 5,000).
	1. A loss of housing for > 30 households in a municipality.
	2. A loss of housing for > 1,000 households per prefecture and 15 households per municipality.
	3. A loss of housing for > 5,000 households in a prefecture.
B	When lives or the well being of a large number of persons has been harmed or when there is a danger of such harm.

2 Disbursement and fiscal responsibilities

A. Prefectural disbursements.
   Relief expenses will be disbursed by the prefecture in which the relief activities are undertaken.

B. National Treasury.
   If the disaster relief expenses disbursed by the prefectures exceeds 1 million yen, the expenses will be born by the national government, see Table 2.4.

Prefectural disbursements/ normal tax revenues	Rate of national fiscal responsibility
2/100 or less	5/100
Between 2/100 and 4/100	8/100
More than 4/100	9/100

Table 2.4 Expenses born by the national government when relief expenses disbursed by the prefecture exceeds 1 million yen.
(July 1993: 1000 Yen = approximately 8 ECU or 9 US$)

3   Content of Relief Activity

1.   Establishment of evacuation areas.
2.   Construction of temporary emergency buildings.
3.   Distribution of water and food, such as boiled rice.
4.   Distribution and lending of clothes, bedding and necessities.
5.   Medical care and midwifery.
6.   Rescue of disaster victims.
7.   Emergency repair of houses.
8.   Supply or rental of funds, tools and materials necessary for a living.
9.   Provision of school supplies.
10.   Burials.
11.   Investigating and handling of corpses.
12.   Removal of obstacles such as rocks, earth and trees.

Nature of Relief Activity	Subjects	Expense Limitations	Period
Evacuation areas	Victims or persons who are in danger of becoming victims	Shelter establishing costs. 100 persons per day up to ¥10.000 special rates for the winter season	Within 7 days of disaster outbreak (extension possible)
Temporary emergency housing	Persons who have lost their homes and do not have the means by which to obtain housing	Average of 23.1 m$_{}$ per house, limit of ¥921.000	Begin construction within 20 days of occurrence of disaster
Boiled rice distribution and other food supplies	Persons who are unable to cook living in evacuation areas or at home	¥720 per person per day. In case of temporary shelter with a relative, etc. (far from disaster site), 3 days worth of expenses payable	Within 7 days of disaster outbreak (extension possible)

Table 2.5 Subjects, expense limitations and periods/relief activity for points 1, 2 and 3 from above **Content of Relief Activity.**

2.3 Systems for Disasters of Extreme Severity

Systems of Designation for Disasters of Extreme Severity

Since the war, Japan has fallen victim to numerous large-scale disasters such as typhoons, every disaster has meant a heavy burden on local finance in the form of huge restoration expenses. Progressive funds from the National Treasury were given according to the degree of disaster, and other financial aid systems already existed at the time.

However, the situation became such that the completion of restoration work was not possible without an increase in the National Treasury funds or rates of financial aid. Thus, with every occurrence of a disaster, the rate of financial aid was inflated somewhat, according to a Special Disaster Exemption Act.

However, the measures imposed by the Special Disaster Exemption Act proved to have a number of draw backs. For instance, they were time-consuming, encouraged lobbying, were partially dependent upon varying circumstances from case to case, and lacked unification. Hence, the need for a comprehensive legal system which applied to all disasters of extreme severity was stressed in conjunction with the rationalization of non-unified and inconsistent National Treasury funds and financial aid systems.

Under such circumstances, the Disaster Countermeasures Basic Act of 1961 reads that to solve the problems of the Special Disaster Exemption Act, the following text should be included in the finance regulations. That a law which provides for special financial aid from the national budget or subsidies for disasters of extreme severity should be established, legislative guidelines were also stipulated.

Disaster Countermeasures Basic Act - Article 97
The government, in the event of the occurrence of a disaster of extreme severity, as separately prescribed by law, should take measures to insure rapid and proper emergency actions or disaster restoration. This will include taking the necessary measures to insure the aptitude of the burden of expenses on local public organizations in which the disaster occurred and to stimulate the enthusiasm for disaster restoration of victims.

The act concerning special financial support to deal with designated disaster of extreme severity (Disaster of Extreme Severity Act) was established in 1962 in response to this article.

Disaster of Extreme Severity Act - Article 2
Upon hearing the opinion of the Central Disaster Prevention Council,
"when a disaster has occurred in which the national economy has been
affected and in which it is necessary to lessen the burden on local
economies or to give special assistance for victims, the concerned
disaster will be designated by ordinance as a disaster of extreme
severity". Furthermore, of the special measures established in this act,
those which should be applied to disasters of extreme severity shall be
designated under the same ordinance.

Designation of a Disaster of Extreme Severity
There are no regulations as to which sort of disasters should be
designated disasters of extreme severity or which measures of which
article should be applied.

This is due to the fact that the conditions and degrees of disasters are
extremely complex and diverse. This means that it is both difficult and
inappropriate to legally establish a criteria which would automatically
judge the type of measures to be applied to the degree of disaster.
Therefore, the legal decisions on what kind of disasters are to be
designated a disaster of extreme severity and what kind of measures
apply, are first to be deliberated by the Central Disaster Prevention
Council and then to be determined individually. Specific judgemental
criteria are to follow the "Criteria for Designation of Disaster of
Extreme Severity" set by the Central Disaster Prevention Council.

Criteria for Designation of Disaster of Extreme Severity

The government designates disasters of extreme severity and applicable
measures based on a criteria for designation established by the Central
Disaster Prevention Council, a government body.

The reason this method was adopted was because it was thought that
while on the one hand, a designation criteria had to be as objective as
possible, it also had to be as flexible as possible in its usage.

The designation criteria were established by the Central Disaster
Prevention Council in December 1962. The present designation criteria
are made up of the following 10 clauses:

1. Disaster restoration works of public utility facilities.
2. Disaster restoration works of agricultural land.
3. Disaster restoration works of agricultural, forestry and fishing.
4. Exemption act for loans for natural disaster rehabilitation.
5. Disaster restoration works for forests.
6. Medium and small-sized enterprise-related exemptions.
7. Disaster restoration works of public social education facilities and private school facilities and infectious disease prevention acts undertaken by the municipality.
8. Construction works of public housing for disaster victims.
9. Small scale disaster bonds.
10. Measures other than 1) through 9).

System for Disasters of Extreme Severity on the Spot
Because a disaster must be of a very large scale and the disaster area must span a wide area to be designated a disaster of extreme severity, it is actually difficult to designate a disaster as such.

However, depending on the condition of a disaster, there are cases where extremely severe damage could be done to a specific area but which would not be considered as serious damage on a national level. For this reason, the Act for Disasters of Extreme Severity on the Spot was established in 1965 stating that if damage within the limits of a municipality exceed a certain level, then exemption measures can be taken for the concerned municipality.

In the designation of a disaster of extreme severity on the spot, areas must be limited, as for instance, "concerning the area of City A" or "concerning the area of Town B or Village C". Furthermore, it is clarified that only exemption measures for municipalities be applied.

Criteria for Designation of Disasters of Extreme Severity
As an example of the criteria for the designation of disasters of extreme severity, disaster restoration works of public utility facilities (1) and assistance for construction works of public housing for disaster victims (8) will be indicated.

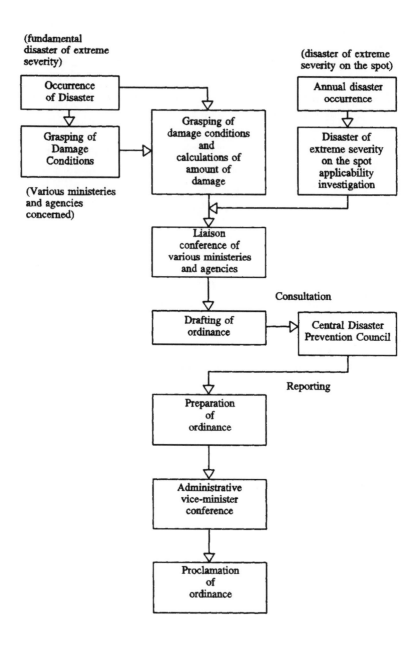

Fig. 2.6 Designation procedure of disaster of extreme severity and measures for application.

Criteria for Designating Special Financial Support Concerning Disaster Restoration Works of Public Utility Facilities
(Either A or B below must apply):

A. Expected assessment value > standard national tax revenues multiplied by 4/100

B. Expected assessment value > standard national tax revenues multiplied by 1.2/100

and

(1) Expected assessment value for the prefecture > at least 1 prefecture with the standard tax revenues for prefectures

or

(2) Expected assessment value for municipalities within the prefecture > at least 1 prefecture where standard tax revenues for municipalities within the prefecture multiplied by 25/100

where the expected assessment value is the expected amount of damage value caused by the disaster and the standard tax revenue is the expected local tax revenue value of local public organizations according to the method set by the local tax grant act.

Exemption Act for Support to Public Housing Construction Works for Disaster Victims

A		Number of lost houses in entire disaster area	>	more than 4,000 houses
B	1)	Number of lost houses in entire disaster area, and	>	more than 2,000 houses
		number of houses lost within a single municipality for more than 1 municipality	>	200 houses or more than 10% of total number of houses
	2)	Total number of lost houses for total area of damage, and	>	more than 1,200 houses
		number of houses lost within a single municipality for more than 1 municipality	>	400 houses or more than 20 % of total number of houses

Table 2.6 Conditions for support to public housing construction works for disaster victims.

Content of relief when a Disaster has been designated as one of Extreme Severity

Financial Aid to Local Public Organizations by the Pooling Calculation Method

Before the Act on Disasters of Extreme Severity took effect, financial aid to local public organizations was given to individual disaster restoration works in the form of increased subsidies. Thus, though highly illogical, if the expenses necessary for disaster restoration works were the same, both wealthy and poor local public organizations received the same amount of financial aid.

Therefore, instead of increasing subsidies for individual disaster restoration works, the present Act on Disasters of Extreme Severity decides on the final amount of financial aid through a progressive calculating method.

As this method takes the total amount of fiscal responsibility of disaster restoration works of a local organization into account and reduces the burden of this total amount, it is called the composite burden reduction method or the pooling calculation method.

According to this method, the more an organization lacks in funds and the bigger the degree of damage, the greater the support from the National Treasury and the more unified and more logical the financial support.

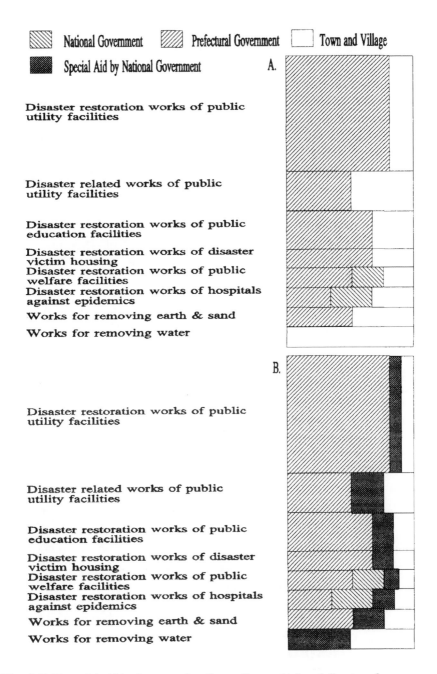

Fig. 2.7 Financial aid in the case of ordinary disaster (A.) and disaster of extreme severity (B.).

2.4 Issues on the present disaster relief system

In the sense that it prevents the bankruptcy of local organizations hit by disasters, our country's Disaster Countermeasure Basic Act is a very effective one. It does, however, embody some problems.

Local self-governing bodies are unable to attend to the complicated procedures for receiving financial support from the national government: The National Land Agency is the overall coordinator through which the government allots aid to the individual ministries and agencies which then implement the aid to matters in their prospective jurisdictions.

However, as individual aid packets span a number of different ministries and agencies, procedures for each one of these ministries and agencies must be adopted.

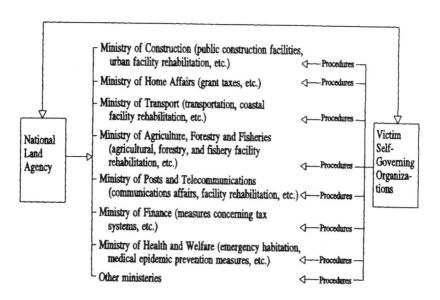

Fig. 2.8 The implementation process for aid to a Victim Self-Governing Organization.

The fiscal responsibility will lie solely in the hands of the self-governing body if it seeks to take actions which do not come under the items for financial support from the national government:
The financial support items of the national government, though improvements slowly continue to be made, are basically still based on the socio-economic conditions of 30 years ago. They often do not apply to the needs of disaster victims of present times.

For instance, in temporary emergency housing construction, when the national government gives financial aid, the standard size of a housing is 23.1 m^2. In Japan a present, this is unrealistic as it does not meet the needs.

Self-governing organizations must be prepared to bear the financial burdens themselves if they want to fulfil the diverse needs of disaster victims.

Adjustments among self-governing bodies in the occurrence of a disaster of extreme severity spanning a wide area can become a major issue:
In the 30 years since the establishment of the Disaster Countermeasure Basic Act and the Act for Disasters of Extreme Severity, there have been no occurrences of large-scale disasters resulting in the conflict of interests between self-governing bodies spanning large areas, namely between prefectures. In the occurrence of a large scale disaster (for instance, a second Great Kanto Earthquake) which would cause a sharp conflict of interests between a number of self-governing bodies on issues such as the procurement of emergency supplies, rehabilitation, restoration planning and procedures, etc., adjustments would be a major issue.

2.5 Role of the Self Defence Forces

Defence Agency as designated administrative bodies

The Self Defence Force is one of the Designated Administrative Bodies, therefore responsible for two obligations:

1. To produce the "Operational Plan for Disaster Prevention" in accordance with Articles 36 and 37 of the Disaster Countermeasures basic Act.

2. To produce "Intensified Plan for Earthquake Disaster Prevention" in accordance with Article 6 of the Large-Scale Earthquake Countermeasures Act.

Large-scale and systematic manoeuvres have been performed every year on September 1st, "The Day of Disaster Prevention".

Disaster dispatching of the Self Defence Force

Act on Self Defence Force
The basic law for disaster dispatching of all troops of the Self Defence Force is the "Act on Self Defence".
There are the following cases in Disaster Dispatching:

- Dispatching troops for disasters

    a. After receiving the request of prefectural governors.
    b. Without waiting for the request of prefectural governors.

Article 83: "In the instance of a disaster, natural or other, where it is deemed necessary to, for the protection of human lives or property, prefectural governors and others designated by government ordinance can make a request for the dispatch of troops to the commissioner or another designated by him".

Troops can be dispatched for disaster relief when the commissioner or one designated by him receives the request of the previous article and deems it necessary. However, in the occurrence of a disaster, natural or other, which calls for particularly urgent aid in which it is understood that the situation cannot await the request of the previous article, troops may be dispatched without awaiting the request (a,b).

When disasters like fire or others occur in or near buildings/facilities of the Self Defence Force, the chief of troops can dispatch the force.

- Dispatching troops for earthquake disaster, (for this case the order comes from the Prime Minister).

Article 83 No. 2: "When the commissioner receives a request according to the same rules by the National Headquarters for

Earthquake Disaster Prevention designated by the Large-Scale Earthquake Countermeasures Act, he can dispatch troops for support".

- Concrete tasks

  Major tasks of troops for reconstruction of disasters and stabilizing social life are as follows:

  1. Catching the situation and lifesaving (the first duty).
  2. Repairing roads and emergency transport of personnel and materials.
  3. Prevention of extension of fire and flood.
  4. Providing foods and water, prevention of epidemics.

2.6 Lifesaving and victim relief activities

Lifesaving and emergency measures for the relief of disaster victims are to be implemented rapidly and accurately by national government organizations, local public organisations and corporations. As each organization fulfils its responsibilities, close mutual contact with the others through laws and ordinances and disaster prevention plans is essential.

Though the contents of these activities cover very wide fields, we can place them into nine categories. Each category has its own responsibility, the subjects of saving, the concrete works for saving and main and sub organisations for works which are fixed by relating laws and regulations.

1. Warnings.
2. Emergency measures for fire fighting, flood and others.
3. Rescuing, saving and protecting victims.
4. Education.
5. Emergency repair.
6. Health and hygiene.
7. Maintaining social system.
8. Securing emergency transport.
9. Other measures.

2.7 Exercise of public authority for the dismantling and handling of victimized buildings

Exercise of public authority for handling of victimized buildings

In a situation where it is necessary to prevent the spread of disaster or for lifesaving, the administration can order disposal by the owner. If the owner will not follow the order, a forced disposal can be conducted.

1. Immediate forced disposal due to urgency.
   There are disposals during or immediately after disasters requiring urgency, the following are examples of such instances.
   - Measures normally assumed as necessary for the prevention of danger may be ordered or executed by another, (Act for Police Duty Performance).
   - Objects which may catch fire are to be disposed of to prevent the spread of fire and so save human lives, (Fire Services Act).

2. Exclusion can be due to unlawful acts or violation of duties concerning rehabilitation and restoration in the stages after a disaster has subsided. Disposals can be ordered when social handling is requested due to danger, toxicity, and obstruction.
   - Emergency public burden (Disaster Countermeasures Basic Act).
   - Orders for the removal of dangerous or toxic facilities (Building Standards Act).
   - Orders and executions by proxy for the removal of unlawful construction on streets and roadsides (Road Traffic Law).
   - Expropriation (Land Expropriation Law).

Duties of the administration in implementing the forced execution and execution by proxy of cases 1 and 2:

It is necessary to limit these to a minimum of the most necessary duties in accordance with the rules of proportion.
- In the Fire Services Act, provisions for a guarantee where destructive fire fighting and so on are implemented.
- Provisions for owner burdens for the duties of guarding and maintenance and other expenses for the subjects of emergency public burdens in the Disaster Countermeasures Basic Act.

Use of public authoritative powers toward victimized buildings which are privately owned

It is necessary to clarify requests, agreements or abandonments by the private user in order to execute the smooth handling of rehabilitation and restoration based on social requests.

- Abandoned objects, those for which ownership rights have been abandoned,
- waste objects, those no longer needed and thus discarded,
- ignored objects, those for which ownership still remains but which are ignored.

The provisions for the abandonment of ownership rights are not clear. There are instances in which the continuation of ownership rights are permitted for waste objects. Owners need to confirm their wills.

Thus, in situations where urgent forced removal is unnecessary and where it is neither unlawful nor in violation of duties to do so, it is extremely difficult to execute public authority for the removal of privately owned victimized buildings.

The present Japanese body of law, in handling of victimized buildings, particularly privately owned ones, makes it necessary to resolve the various conditions of the buildings at present such as whether they are guaranteed or mortgaged articles. However, it is predicted that a great amount of time would be spent on doing this and where urgency is required, countermeasures such as "authorization by the national government" become necessary.

As an example of the confirmation of an owner's will in handling victimized buildings is the instance of the Great Fire of Sakata. In this case, the local "City Bulletin" was used to urge questions and the confirmation of the will of owners. If there was any inconvenience, it was to be voiced. No response was interpreted as an owner in agreement. This was done according to the civil law where "regarded authorization" was based on the "declaration of intent by public notification".

2.8 Rehabilitation of lifelines

Lifeline related institutions

Lifeline related institutions are either Designated Local Administrative Bodies or Designated Public Corporations. They engage in emergency measures in close cooperation with the Prefectural Disaster Countermeasures Headquarters in times of disasters.

Water and Sewage	-	Under direct supervision of the prefecture
Gas	-	Tokyo Gas Co. (DPC: Designated Public Corporation)
Electricity	-	Tokyo Electric Power Co. (DPC)
Communications	-	NTT Co. (DPC)
Broadcasting	-	NHK Co. (DPC)
	-	10 private broadcasting stations (DPC)
Roads	-	Under direct supervision of the prefecture for roads, forestry for tracks and city for city roads
	-	National roads come under the supervision of the Kanto Region Construction Bureau of the Ministry of Construction (DPC) Highways come under the supervision of
	-	the Japan Highway Public Corporation or Tokyo Expressway Public Coporation (DPC)
Railways	-	The JR (DPC)
	-	The 9 private railway companies
Coast	-	Second Coastal Bureau of the Ministry of Transport or the Kanto Transport Bureau (DPC)
Rivers	-	Kanto Region Construction Bureau of the Ministry of Construction (DPC)

Fig. 2.9 Table showing the Tokyo area disaster plan for Earthquakes.

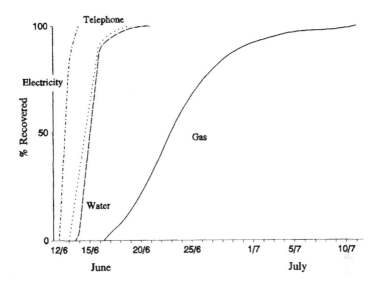

Fig. 2.10 Recovery of lifelines in Sendai City. (Miyagi Prefecture Offshore Earthquake, 1978).

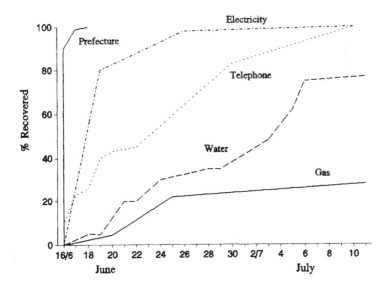

Fig. 2.11 Recovery of lifelines in Niigata (Niigata Earthquake, 1969).

Confusion in lifeline rehabilitation

An example of a modern city being struck by an earthquake in our country, resulting in a big issue concerning lifeline rehabilitation, is the Miyagi Prefecture Offshore Earthquake.

Individual businesses plan their own rehabilitation plans separately. What became clear from this disaster was that the confusion for manpower and equipment and the difficulty in making adjustments between the lifelines would obstruct rapid rehabilitation if an earthquake of even larger scale were to hit an area such as Tokyo.

As electricity is supplied by overhead wire in our country, the rehabilitation of electricity would be the fastest whereas gas would be the slowest. Aside from the examination and restoration of the underground pipework, the major reason for the delay in the recovery of the gas supply is that the main valves to all houses must be examined as a safety measure before the gas can be turned on.

2.9 Cases of disaster relief and restoration activities

The Great Fire of Sakata

Conditions of the great fire

Time and place	-	29th October 1976, 17.40 in a wooden structure movie theatre
Extinguishing	-	30th October 1976, 05.00 (after 11 hours)
Damages	-	1 dead, 964 injured (of which 10 were seriously injured)
		1,774 fire destroyed buildings
		1,123 victimized households
		3,300 victims
		22.5 hectares of fire destroyed area (800m east-west, 300m north-south)
		40.5 billion yen in damages
Weather Conditions	-	Rain, winds averaging 12.2 m/s (maximum 22.7 m/s)

Conditions for the disposal of rubble

Dispatch of Self Defence Forces:
The Self Defence Forces arrived whilst the fire continued to spread between midnight of the 29th and the early morning of the 30th. After the fire was extinguished, the removal of rubble from the roads commenced as the securing of road traffic in the disaster area was the first objective.

The disposal of rubble in the general city area was conducted between November 6th to the 15th, as a municipal consignment task. This task was completed in a week, and some of the facts related to this effect are listed below.
- A total of 1,334 motor vehicles and 12,500 people were mobilized. The amount of rubble handled totalled 23,000 cubic meters.
- It was argued that the primary work of the SDF is the saving of human lives and not the disposal of garbage. Nevertheless, the contribution of the SDF's activity was no small effort.
- Not only the SDF but local construction companies were also engaged in the disposal activity.

Selection of the disposal area:
Securing a disposal area was a very difficult issue. At first, coastal accumulation and ocean dumping were considered. However, these were met with strong opposition from the prefectural coastal controllers and thus abandoned. Fortunately, an industrial development complex was available as a temporary accumulation area.

An agreement was reached on employing the area not as an ultimate disposal area but at a temporary one, under the condition that it would be cleared as a later date. There were claims of the area being too far away, but at any rate, it was secured.

Even more fortunate was that the remains of a river gravel pit were eventually located as the final disposal site. The Gekko River had numerous large holes from gravel extraction, and was designated a danger area, since children were falling in the holes and drowning. The rubble and garbage which was temporarily lying in the industrial complex was therefore used to fill these holes. The number of motor vehicles used to carry the estimated $73,000m^3$ of rubble and garbage was 2,765.

If it were not for those disused gravel pits, disposal would have been a serious problem, and the waste would have had to be carried to a mountain site or used in the construction of a dam to be constructed in a valley.

Designation of garbage in privately owned areas:
At first the city had decided that garbage disposal in private areas was to be conducted by the owners. However, due to the strong request for collective disposal, it was decided that the city would carry out this task. The problem was what to do about the materials from privately owned disaster-struck areas.

The city asked those who did not wish to have their garbage disposed of to make themselves heard through the city bulletin. Where there was response, the material were designated as garbage (regarded designation) and disposal was conducted.

- Many objects such as half destroyed buildings and unmovable safes left on completely destroyed premises emerged.
- Problems such as distinguishing completely destroyed buildings from half destroyed ones and whether or not to wreck a building emerged.
- However, it was impossible to confirm the wishes of each and every one of 1,000 households.
- Two or three troublesome cases emerged later but these did not grow into bigger problems.

Miyagi Prefecture Offshore Earthquake

Conditions of earthquake damage

Time and place - June 12, 1978, 17.14
Magnitude 7.4
Epicentre in the ocean floor 120 km east of Sendai City.

Damages to - 13 dead, 9,300 injured
Sendai City 4,250 completely and half destroyed buildings
74,000 partially destroyed buildings.

Characteristics of earthquake damage

Sendai City produced a study report in December 1979, a year after the earthquake. The following are the characteristics of the damage mentioned in the report:

1. The majority of deaths were due to the collapse of concrete fences. Elderly persons and children comprised the majority of such victims. Elderly persons, women and children were also the most injured. Many of the injuries occurred indoors.
2. Damage to buildings was regional. Concentrations were seen in alluvial plains and in developments on large reclaimed marshlands found in hilly areas. Furthermore, old houses where the most exposed houses to damages.
3. In comparison to the severity of the earthquake, the damage suffered by humans was not large. The occurrence of secondary and tertiary disasters, which are the causes of such disaster spread was controlled. Thus, panic due to these did not occur.
4. Despite the halt in the supplies from urban lifelines: water, electricity and gas, no panic occurred. Responses by citizens was cool.
5. General restoration was a smooth process though the complete rehabilitation of prefectural gas took nearly a month.

Survey results from the study

1. Occurrences of injuries and illnesses:
    - 345 occurrences (rate of occurrence 1.74%)
    - of the occurrences of injuries 77.4 % were surgical.
2. Treatment of injuries and illnesses (of the 345 victimized):
    - self administrated 60.0 %
    - professional treatment the same day 13.0 %
    - professional treatment the next day 20.9 %
    - not known 6.4 %

The number of persons who had received professional treatment was 117, of which 5 (3.4%) were transported by ambulance.

3. Place of occurrence of injuries and illnesses (of the 345 victimized):
    - 55.9 % at home, 14.8 % in other buildings, 9.9 % out of doors, 19.7 % not known.

Conditions of garbage and rubble disposal

	Household Garbage (direct collection by prefecture) [tons]	Garbage produced in the workplace [tons]	Total [tons]
Incinerated	423	119	542 (77.5%)
Reclaimed	5	152	157 (22.5%)
Total	428	271	699 (100%)

Table 2.12 The above table represents the garbage produced daily in Sendai City before the earthquake disaster.

Two days after the earthquake and onward, large amounts of rubble were produced in Sendai City and a piece of reclaimed land free to citizens was made available for the disposal of the garbage. The authorities also used all their power to collect and move garbage in the city. Nevertheless, a large amount of garbage was left accumulated in the city the disposal of which was completed on June 21st, 10 days after the earthquake.

The figures below relate to the amounts of garbage moved to the reclaimed land site after the earthquake disaster (June):

Directly collected by the city	11,324 tons
Carried in by licensed companies	2,834 tons
Carried in by construction companies	1,824 tons
Carried in by citizens	13,801 tons
Total	29,783 tons

The total amount of 29,783 tons of garbage represents 49 days worth of garbage of Sendai City before the earthquake.

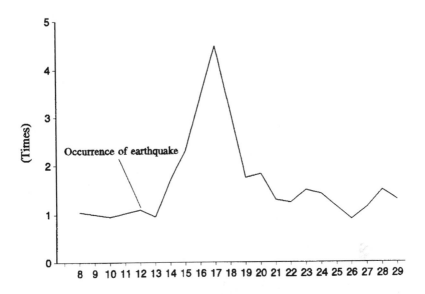

Fig. 2.13 Amount of garbage in June 1978, Sendai City.

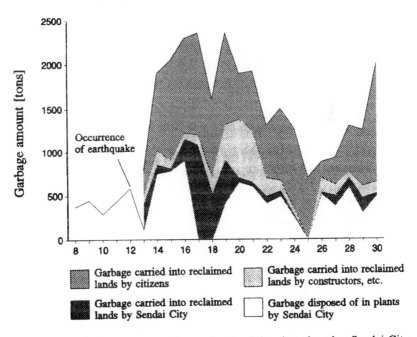

Fig. 2.14 Change in the amounts garbage after the Miyagi earthquake, Sendai City, 1978.

2.10 Study of countermeasures required for the next Great Kanto Earthquake

The south Kanto region including Tokyo was subjected to disastrous damage during the Kanto Great Earthquake in 1923, (magnitude 7.9).

Total damage:
- 142,800 persons dead or missing
  103,700 persons injured
- 128,300 destroyed (broken) houses
  447,100 destroyed (burnt) houses

Damage to the city of Tokyo:
- 60,100 persons dead or missing
  15,700 persons injured
- 3,900 destroyed (broken) houses
  366,300 destroyed (burnt) houses
  (44 % of total area of city destroyed by fire).

Aspects of earthquake damage in Tokyo
There are two major aspects of earthquake damage in Tokyo.

The first pertains to damage done mainly to the ground, buildings and facilities. This will be called earthquake subject damage. For damage of this type, the big issue is rapid rehabilitation, particularly the recovery of urban functions centring on the recovery of lifelines.

The other can be called fire subject damage when large areas are destroyed by fire. In the non-urban areas of south Kanto (the 4 prefectures of Tokyo, Kanagawa, Saitama and Chiba), there is a danger for this type of damage. Assuming that the second Great Kanto Earthquake will strike on a winter night, it has been calculated that 800 km^2 of land will be destroyed (National Land Agency, 1988).

Issue of planned urban restoration - a case of destruction of a large area of land in South Kanto
At present there is no premeditated plan for occurrence of such a situation. We are still at the stage where whether or not urban restoration plans should be established ahead of time is just beginning to be examined.

Not just those of a number of cities but also the agreement of all four prefectural governors and furthermore, of the national and self governing organisations are necessary for planned urban restoration in the aftermath of a large-scale earthquake disaster.

Solidification of a restoration ideology and consensus formation among citizens cannot be expected directly after an earthquake when no preparations have been made. When a large-scale earthquake indeed strikes, it is highly possible that a disorderly city will be built again.

Cleaning and prevention of epidemics (Tokyo area disaster plan - earthquakes

The following are details of the cleaning and prevention of epidemics plan for the Tokyo area for any future earthquakes published by the Tokyo Metropolitan Government.

Garbage Disposal Plan

The prefecture of Tokyo predicts the amount of garbage produced following the occurrence of an earthquake as follows:

Garbage from collapsed wooden houses	958.3	thousand tons
Garbage from fire destroyed wooden houses	1,112.2	thousand tons
Garbage from water disasters	14.1	thousand tons
Garbage from general usage	128.5	thousand tons
Total	2,213.1	thousand tons

Of the 6 million tons of garbage which the prefecture of Tokyo collects annually, 2 million tons are reclaimed. Earthquakes are the cause for 1/3 of the annual garbage produced. It is proposed that this garbage be handled in the following way:

1. First Countermeasure - As it is difficult to carry garbage into a disposal area, a temporary area for leaving the garbage is secured. Public spaces free of environmental sanitation problems are preferable. This garbage must then be collected within 10 days after the collection of garbage has become possible.

2. Second Countermeasure - The garbage carried into the temporary holding area must be brought to the disposal area within 20 days after the completion of the first countermeasure.

2.11 Dismantling, rubble disposal and recycling of victimized structures

This issue remains unexamined in our country and its importance is only just beginning to be considered. In the following it is hypothesized that a large number of reinforced concrete (RC) structures have been damaged by a major earthquake. Measures for dealing with the catastrophe and various emerging problems will be discussed.

Assessment of damage to victimized structures

The damage is assessed by the following procedures:

1. Preparing a standard for evaluation of damage.
Based on past experiences of the effect of earthquakes on RC and other structures, a standard for assessing the degree of damage is prepared. The preparation of the "classification of assessment of damage for reinforced concrete structures" is being led by the Ministry of Construction in Japan.

2. Training for classification and assessment of damage.
Education for assessing the degree of damage; namely for first class building works execution managers and first class civil engineering works execution managers or technicians with evaluating abilities should be implemented not after a large-scale disaster has occurred but in normal times.

3 Evaluating damage to see whether a structure should be maintained or dismantled.
Evaluations, as a rule, are done by visual inspection. The basis for evaluating are roughly as follows:

1. Can be used in its present state for a long period of time (with slight repairs).
2. Needs a certain amount of repair but can be renovated. This building can be used as it is for the housing of disaster victims for a while.
3. No immediate danger of collapsing at any moment, but should be designated as a "no entry" area. It will eventually be dismantled.
4. To be immediately dismantled due to danger of collapsing.

Dismantling victimized structures and the problems involved

In Japan techniques incorporating large size concrete breakers are mainly used for dismantling RC structures. For sturdy structures large size hammer breakers and combinations of different demolition techniques are used. For underground structures a giant hammer is employed. Demolition methods and techniques are presented in chapter 4 and described in details in the reports of RILEM TC-34-DRC and TC-121-DRG, and the proceeding of the second and third International RILEM Symposia on Demolition and Reuse of Concrete and Masonry. The dismantling and disposal of the large number of damaged structures resulting from large-scale disasters such as earthquakes generates a number of problems:

1. Development of equipment for the discovery and confirmation of survivors.
The development of equipment for identifying and locating survivors (eg. detectors of audio equipment, fibrescopes, carbonic acid) is necessary.

2. Dismantling and removal techniques for the rescue of survivors.
The rapid rescue of victims trapped inside collapsed structures without injuring them is important. The combined use of effective demolition techniques such as large size hammer breaker and breaker, along with cranes and bulldozers is necessary.

3. *Development of rapid dismantling and removal techniques of collapsed buildings.*

RC structures and other concrete structures are extremely sturdy due to their earthquake endurance design. If these structures were to collapse in large numbers, it would become necessary to rapidly dismantle and remove them.

With regard to the dismantling technique, noise, vibration and falling objects are not a problem, thus the use of explosives and/or the chain and ball would be the preferred techniques.

As the chain and ball is not a technique used in Japan at present, we are forced to start by looking for them after a disaster strikes. The use of explosives should be left to specialists.

Aside from these, large size hammer breakers, breakers, crawler cranes and bulldozers must also be used effectively.

Issues on rubble disposal and recycling

In Japan, approximately 40% of construction waste in the Tokyo area is disposed of by reclaiming land. Of this, approximately 45% of concrete waste is recycled into road base aggregate. Recently, the garbage produced in Tokyo was rejected from nearby prefectures, this is a serious situation, and at present the Kanto region can only take 6 months worth of additional waste. In October 1991, the Ministry of Construction established a Recycling Act and defined the waste products of building construction as "by-products". To particularly promote the recycling of concrete, asphalt concrete, used timber, and other debris, these were called "designated by-products" which were ultimately to be 100% recycled.

There is the possibility that large amounts of dismantled rubble produced by a major earthquake in Tokyo will not find a disposal area and furthermore be rejected by other prefectures. The method for disposing concrete should be considered from this kind of a perspective.

1. *To establish a temporary accumulation area for the use of a crushing machine to promote recycling.*

The problem here is securing such an accumulation site. Since Tokyo is full of parks and publicly owned land, it is necessary to designate land for such use from normal times.

2. *Boosting the level.*

Tokyo has a vast area of urban lowland, and there are suggestions to utilize the construction waste to raise the level of this land. Despite the urgency of time directly after a disaster, it is still necessary to make careful plans and have the agreement of the residents for using rubble carried in from other areas for this objective.

3. *Accumulating the rubble evenly on disaster-struck land.*

Japan's RC structures generally use 0.6 m^3 of concrete per 1 m^2 of floor area. Thus, if the ratio of building area to ground area of a 5 story RC structure were 60 %, completely dismantled, this would mean approximately $5 \times 0.6 \times 0.6 = 1.8 \text{ m}^3/\text{m}^2$ of accumulated concrete. If the real accumulation rate of concrete were settled at 0.75, this would mean a 2.4 m increase in the level of the land.

In this way, the amount of rubble produced from a disaster-struck building should be judged by the amount of concrete in a structure within individual urban blocks in normal times. Furthermore, individual buildings which would be damaged at different levels on the seismic scale should be hypothesized and the amount produced per urban block needs to be predicted.

If the ground were to be boosted in such a way, lifelines such as those for gas and water would be buried under the roads. Thus, the roads need not be raised. Instead, a concrete box the width of the road and of the necessary height could be placed on the road, making the buildings approachable, thus service tunnels, for such things as electricity and sewage, could be secured.

4. *Accumulating rubble at fixed widths along embankments.*
The Tokyo area has long embankments with thin walls in need of strengthening. Large amounts of rubble are taken and accumulated along embankments at considerable widths. Those who own the land on which this is done are in turn contributed the land on top of the embankments after the walls are made wider.

5. *When damage is extraordinarily major.*
When it is difficult to dismantle victimized structures or when due to the eruption of a volcano a large amount of volcanic ash has accumulated, activity should be moved to another location.

Issue of recycling concrete

There are many problems concerning the reuse of dismantled concrete material as concrete aggregate. In Japan there is not yet an example of an actual structure made with recycled aggregate concrete. The issue of reusing the recycled materials from a collapsed RC structure after a large-scale disaster is to be examined.

1. *Using concrete crushing factories of regions unaffected by disasters.*
In the Kanto region, centred around Tokyo, there are numerous small crushing plants. Dismantled concrete is carried to unaffected crushing plants and recycled aggregate material is carried back to the site on the return trip, using it for rehabilitation construction.

2. *Establish numerous concrete accumulation areas in disaster areas.*
Move the existing crushing plant facilities, produce recycled aggregate materials, and reuse these.

3. *Construction of RC structures using recycled aggregate materials.*
It is necessary to hypothesize on the quality of the aggregate made from recycled concrete and to conduct tasks to measure the properties (e.g. strength + shrinkage) of concretes made with this material. Furthermore, it is necessary to actually use the recycled aggregate material in RC structures and undertake long term tests on its performance.

3

DAMAGE ASSESSMENT AND CLASSIFICATION

Johan Vyncke and Hans Geenens, Belgian Building Research Institute
Dr Susanne Buchner, Gifford & Partners Ltd, England

3.1 Introduction

After a large scale disaster, as in an earthquake, many relief activities such as saving of lives, medical care, prevention of the spread of fire, transportation of habitants and relief personnel, food and water supply, and distribution of clothes, are developed. For this, the rehabilitation of the infrastructure of the area is very important, and should occur as fast as possible; blocked roads should be opened, emergency bridges installed, bridges repaired, etc. Another main concern after such a damaging disaster is housing. Many houses are likely to be damaged or have collapsed, leaving homeless people searching for shelters. Uncontrolled reuse of dwellings can augment the number of victims, due to building collapse caused by aftershocks or live-loads, and can obstruct certain relief activities. To avoid the uncontrolled reuse of buildings it is very important to examine the buildings as soon and as fast as possible and to prohibit the entry to hazardous dwellings. Thus, whether it concerns buildings or infrastructure, a quick damage assessment after a disaster is important in order to achieve an effective relief campaign and to make a fast rehabilitation of economic and social life possible. This, however, needs an effective damage assessment plan in the post-disaster period, as an important part of the whole disaster emergency plan.

A quick assessment of the damage is not only very important from an emergency and safety point of view. The gathered information can also help in making difficult political decisions after a disaster. Estimation of the actual damage to structures and its quantification in monetary terms is motivated by the need to plan an effective and timely relief campaign in the short term (days, weeks) and a reasoned and well-grounded policy to allocate funds for reconstruction in the long run (months, years). The following goals for the assessment of structural damage after a violent disaster should be aimed for:

- Prevent the reuse of severely damaged buildings to avoid the increase of the number of dead and injured people, caused by the collapse of damaged buildings due to live-loads or after-shocks. This is indeed the main goal.
- The preparation of quick repair guidelines to enable the structure to be placed back in service, providing the population with safe shelters.
- The damage assessment is an important tool to be used in decisions about repair, demolition, and reconstruction.
- Damage estimation and quantification are needed for decisions about financial relief and financial responsibility.
- The assessment campaign can help to give an idea of the amount and kind of building waste. This is necessary for developing a policy for site cleaning, recycling and reconstruction.
- Statistical processing of the collected data can help to develop new design guidelines and design methods.

Earthquakes are one of nature's greatest hazards on this planet. Many countries of the world situated along the seismic belts, including Japan and the United States, often deal with earthquake disasters. That is the reason why the great majority of the publications concerning damage assessment methods concentrate on post earthquake relief and are of Japanese and American origin.

The best English publication we came across concerning damage assessment and classification in the post earthquake period, is the ATC-20 report [1], an American publication of the Applied Technology Council. Much of what follows has been taken from this report. There are also many interesting Japanese publications about disaster relief following earthquakes but unfortunately they are seldom translated and difficult to obtain. However, thanks to the efforts of Mr. Yoshio Kasai from the Nihon University in Japan, we are able to present some interesting Japanese guidelines concerning damage assessment and classification.

This chapter contains an overview of the basic subjects, concerning damage assessment and classification in post disaster conditions, which we believe are to be considered when developing a practical, damage assessment guide for damage surveyors in the field. Such a practical guide should be elaborated on an international level, as in the emergency relief phase teams from all over the world often collaborate. The general guidelines should of course allow for different local conditions as described in chapter two concerning disaster countermeasures in Japan. Important local factors that should be mentioned are: design methods, construction materials, climate, political situation, and ethnic, cultural and economic conditions.

The list below outlines the different subjects described in this chapter.

1. Introduction

2. Evaluation strategy

The major problem for a fast damage assessment after a disaster is the lack of trained, experienced structural engineers. Even with many volunteers and the assistance from outside there is more work than can be handled. A good strategy that takes account for this is based on a three level evaluation: Rapid, Detailed and Engineering Evaluation.

3. Survey forms and manuals

A suitable survey form and an efficient manual can help to make a fast and reliable assessment in the field. While developing such a form and manual there are many different elements that should be taken account of.

4. Classification of damaged structures

After the structural elements have been assessed locally, the construction should be classified as a whole. This classification will be a major index for decisions in the relief, rehabilitation and reconstruction period.

5. Posting of the building

It is very important in the relief period to avoid further victims caused by building collapse due to after-shocks or the uncontrolled reuse of damaged buildings. Very important tools for doing this are the posting of assessed buildings and barricading of dangerous areas.

6. Damage assessment of buildings

Since the technical expertise of the inspectors will inevitably cover a broad range, damage assessment should be done using a unified method. This assessment method will be summarized in the survey manual that will accompany the inspector in the field. Some examples we came across in the literature are presented.

7. Essential facilities, infrastructure and lifelines

Hospitals, police and fire stations, roads and bridges, water, gas and electricity supply, are very important for the rehabilitation process in the destroyed area. Therefore, they should get special treatment in the post disaster period. The assessment should get first priority, and it would be desirable to have a specially trained engineering team to do the inspection. After the assessment the structures should (if possible) be repaired as fast as possible.

8. Organisation

The demand for outside assistance will be high making good co-operation between local, national and international authorities and organisations necessary. Because many different activities interfere in the emergency period, efficient co-operation and an interchange of information between the different activities and professions are also needed. The setting up of an international intervention team of construction engineers might be considered.

9. Recruiting and training of staff

To deal with the lack of staff in the relief period, volunteers should be recruited and trained in the pre disaster period. It is necessary to make the assessment staff familiar with the assessment procedure and prepare them for post disaster conditions.

3.2 Evaluation strategy

General

The introduction presents different goals for the assessment and classification of buildings damaged by a disaster. The most urgent goal for building assessment and classification is to keep people out of severely damaged buildings and dangerous areas. Therefore it is important that the condition of the buildings is examined as fast as possible. However, the major problem for a quick damage assessment after a disaster is that trained, experienced manpower are likely to be in short supply. Even with an influx of assistance from outside the area and the help of many volunteers, there is normally much more work than can be handled by the available staff. Strategies for the safety evaluation of structures need to account for this.

A good strategy, considering the need for judicious use of the limited number of trained structural engineers, is developed in the ATC-20 report [1] and is based on a three levels building safety evaluation:

1. Rapid evaluation: This is normally the first level of evaluation and its aim is to make a quick safety and usability evaluation. It is designed to quickly, and with the minimum of manpower, designate the apparently safe and the obviously unsafe structures. Doubtful structures are designated for a more detailed visual evaluation in a second examination round. The Rapid Evaluation uses less qualified and experienced inspectors and volunteers, who in the best conditions are trained in the pre disaster period. The procedure is simple and clear and is presented in more detail in the following section.

2. Detailed evaluation: This second level is a more in depth evaluation method whose goal it is to examine the structure in detail and make a damage classification. It is used to evaluate the damage of buildings whose conditions are still doubtful after the Rapid Evaluation. It consists of a thorough visual examination of the entire building, inside and out, particular its structural system. This evaluation procedure collects more information about the building, that can be used for later decisions concerning repair, demolition and reconstruction. A summary of the procedure is presented on page 57.

3. Engineering evaluation: Whenever a building has been damaged to such an extent that it is not possible to use visual inspection techniques alone an Engineering Evaluation is required. This is the third and most thorough evaluation level. The Engineering Evaluation uses more sophisticated, partly destructive, examination techniques and must be ordered by the owner of the building. These are two reasons why we consider the Engineering Evaluation as not being part of the real emergency relief activity. The procedure is summarized on page 59.

Most other publications do not propose different levels of evaluation. They immediately consider a more detailed damage assessment, collecting different kinds of information. Their assessment guidelines are more detailed than those of the Rapid Evaluation procedure, and can only be used by experienced structural engineers. These evaluation methods are comparable with the second evaluation level proposed above. A detailed description of different second level evaluation methods is given later in this work.

However, a three levels strategy that makes good use of the available manpower and fulfils the different tasks of a post disaster emergency operation, seems necessary. Immediately using a more detailed evaluation procedure would take too long, and therefore would not achieve the main goals. For example, after the earthquake in El Asnam (Algeria) of October 1980 [10] an in depth examination of 6,538 structures was made by 100 engineers over a period of two months. Meanwhile many badly damaged structures were already being reused and slums were formed, causing many problems to later demolition works.

Rapid evaluation

Rapid Evaluation is an emergency, safety and usability evaluation method. Its aim is to maximize the number of inspections in the immediate postevent period to avoid more victims due to uncontrolled reuse of damaged buildings. The Rapid Evaluation is performed by evaluating the building for six screening criteria (Fig. 3.1) that are primarily externally observable conditions.

	Condition	Classification
1.	Building has collapsed, partially collapsed, or moved off its foundation.	UNSAFE
2.	Building or any storey is significantly out of plumb.	UNSAFE
3.	Obvious severe damage to primary structural members, severe racking of walls, or other signs of severe distress present.	UNSAFE
4.	Obvious parapet, chimney, or other falling hazard present.	AREA UNSAFE
5.	Large fissures in ground, massive ground movement, or slope displacement present.	UNSAFE
6.	Other hazard present (e.g., toxic spill, asbestos contamination, broken gas line, fallen power line)	UNSAFE or AREA UNSAFE

Fig 3.1 : Basic Rapid Evaluation Criteria [1]

This is normally the first level of evaluation and is designed to quickly, and with the minimum of manpower, designate the apparently safe (posted inspected) and the obviously unsafe structures (posted unsafe). Doubtful structures (posted limited entry) are designated for a more detailed visual evaluation in a second examination round. This evaluation method is designed for use by building inspectors with a basic familarity with building construction. The inspection procedure (Fig. 3.2) takes about 10 to 20 minutes per building.

1.	Examine the entire outside.
2.	Examine the ground.
3.	Enter a building only if the structure cannot be viewed sufficiently from the outside, do not enter obviously unsafe structures.
4.	Post every entrance of the building for limited entry or unsafe, mark if interior as well as exterior were evaluated.
5.	Explain the significance of unsafe or limited entry to building occupants and advise them to leave immediately.
6.	Complete the six criteria checklist on the survey form.

Fig 3.2 : Rapid Evaluation inspection procedure [1]

Detailed evaluation

This second level of evaluation is primarily used to evaluate the safety of buildings posted Limited Entry after the Rapid Evaluation. It is designed to result in the rating of all structures as either safe for use (posted inspected), potentially dangerous (posted limited entry), or unsafe (posted unsafe). It consists of a thorough visual examination of the entire building, inside and out, particularly its structural system. A minimum requirement in the evaluation is that the structure must be capable of withstanding at least a repetition of the event.

Eight basic evaluation criteria are used: overall damage, vertical load system, lateral load system, p-delta effects, degradation of the structural system, falling hazards, slope or foundation distress and other hazards. More specific and detailed guidelines are given for wood frame, masonry, tilt-up, concrete and steel frame structures, and geotechnical and non structural hazards. A thorough description of this Detailed Evaluation sytem is given in section 3.6.

Ideally, the Detailed Evaluation is performed by at least two structural engineers with post event inspection experience and knowledge of earthquake effects on buildings. An alternative is the use of a team consisting of a structural engineer and a building inspector. The Detailed Evaluation procedure (Fig. 3.3) takes 1 to 4 hours per building.

1.	Survey from outside.
2.	Examine the site for geotechnical hazards.
3.	Inspect the structural system from inside the building.
4.	Inspect for non structural hazards.
5.	Inspect for other hazards like elevators, stairs, fire protection equipment, stored chemicals, etc.
6.	Complete the checklist on the survey form and post every entrance to the building.
7.	Explain the significance of unsafe and limited entry posting to building occupants and advise them to leave immediately.

Fig 3.3 : Detailed Evaluation prodcedure [1]

It is of great importance in solving the problem of emergency dwellings to develop criteria for emergency repairs and strengthening (very often cheap and quick bracing is sufficient to make a building safe). After the Rapid and the Detailed evaluation, these criteria could help making the emergency repairs, to rapidly provide the population with shelters and emergency housing. At the same time the surveyors should point out the buildings in danger of collapse. These buildings should be demolished as fast as possible because they are a constant threat for people and dwellings in the area. For both activities good communication between the different teams and professions is required and should be organised in the pre disaster period.

Engineering evaluation

Whenever a building has been damaged to such an extent that it is not possible to use visual inspection techniques alone to assess its safety an Engineering Evaluation is required. This would normaly be done by a structural engineering consultant who may need to remove portions of the building to complete the examination. The decision to conduct an Engineering Evaluation is the responsibility of the building owner.This is the third and most thorough evaluation level. Such a study would typically include detailed reconnaisance and mapping of the damage, new structural calculations, and a quantitative assessment of the strength of the damaged structure. It may take 1 to 7 days or more per building. After this examination, all buildings may be posted inspected or unsafe.

It is important to reiterate that the Rapid Evaluation should not be considered as part of the emergency relief activities. Though, it will be the main tool in making decisions for repair in the rehabilitation period. Once the assessment is made, a decision has to be taken whether to repair or whether to demolish and replace the structure. This decision is mainly based on a compromise between technical possibilities and economical costs, taking into consideration future developments. But there are also policy questions that go beyond the specific technical problems of a single damaged structure. What criteria should be used to place older, structurally damaged buildings back into service ? Should they be restored to their original condition, in which case they may be greatly under strength in terms of the toughness and ductility required for new structures, or should they be strengthened to greater levels ?

Concerning this assessment level it is interesting to refer to the CEB report "Diagnosis and Assessment of Concrete Structures" [8]. In this state of the art report an overview is given of different aspects of the subject and the principal assessment techniques (structural, physical, chemical and electrochemical tests). It is important to realize that practically all tests, that can be applied, are indirect tests that require a professional interpretation by the assessing engineer before he is able to make use of the test results. Therefore the assessing engineer must have a thorough understanding of all his test procedures and the reliability of the results obtained.

Although details of the procedure of investigation will depend on the requirements of the client in addition to the technical needs for

undertaking the work, the overall assessment procedure can be
considered as a six stage process (Fig. 3.4).

Stage	Objectives	Operations
Preplanning	To ensure that the survey is undertaken efficiently with background knowledge available.	Collect all historical data and test results. Standardise reporting forms.
Global scan	To understand the behaviour of the structure. To select areas for detailed examination. To determine optimum measurements techniques.	Visual inspections Photographic record. Non-destructive tests. Selected samplings.
Detailed examination	To provide sufficient, reliable data to enable the condition of the structure to be assessed with confidence.	Load tests. Non-destructive tests. Physical and chemical tests.
Presentation of results	To enable the survey results to be easily assessed and compared.	Computer plotting. Statistical analysis.
Interpretation of results	To use the recorded results to assess the current and future performance of the structure in relation to the requirements.	Structural analysis. Deterioration analysis. Past experience.
Recommendations	To determine what further actions are necessary either for repair, strengthening, preventative treatment or additional surveys.	

Fig 3.4 : The CEB Engineering Evaluation procedure [8]

3.3 Survey forms and manuals

General

In order to make sound decisions based on the damage estimation, two
major requirements are needed: quickness and reliability. Although
Rapid and Detailed Evaluations have a different aim and use a different
skilled assessment team, this can for both assessment levels be met by
the use of suitable survey forms and efficient manuals.

Survey form

The survey form is a normalised questionnaire that has to be filled in by the damage assessment team. The form should be developed to make the assessment faster and to collect all the information required for later decisions concerning short term relief and long term policy. Statistical processing of the collected data at a later stage can help in the development of new design guidelines and design methods. To develop an appropriate survey form it should fulfil the following tasks. The last two will be achieved using an appropriate manual.

- To be practical, the form should be compact and conveniently organized.
- To achieve a quick assessment, the requested information should be kept to a minimum and the information should be directly observable.
- The form should give all the information required for future decisions.
- To make sound decisions the collected data should be reliable and objective.
- If data processing with computer is planned the form could be devised in such a way that processing the data becomes easier.

In developing the form it is important to decide how much and which information is required. Therefore, a compromise between the second and the third tasks is necessary. The more detailed the analysis (and the more information we take into account) the more reliable may be the results. However, a greater number of parameters to be considered during the survey, involves a lot more time and expenditure. The choice of parameters depends on the objectives of the assessment. So, for example, the Rapid Evaluation will need an easier and a more compact form than the Detailed Evaluation.

The basic information required in a damage survey form for the Detailed Evaluation level can be divided into six groups concerning the identification of the survey team (1), the identification of the building (2), the composition of the building (3), the use of the building (4), the structural characteristics of the building (5) and the damage assessment and classification results (6).

Extra information can be obtained by taking photographs (7) of the damaged building from different points. Referring to Petrovski [4], at least one should be taken of a safe building and at least two of an unsafe or doubtful building. For bigger and more complex buildings taking more photographs would provide a more realistic representation. The photographs should ensure the best view of the general condition of the building, the characteristic damage, as well as the posting placard or mark. The numbers of these photographs should be noted on the form.

A compilation of the forms in the literature, presented in Figure 3.5, gives a good overview of the required information. The numbers from the table correspond with the groups of information mentioned above.

1.	Identification number of the survey team Survey date
2.	Identification number of the building Address of the building Owner of the building Number of residents Building age Number of attached buildings and their identification numbers
3.	Gross area Average height Number of stories Basement, yes or no Number of apartments and rooms
4.	Primary occupancy (dwelling, office, commercial, school, government, other) Productive activities
5.	Type of vertical structure Type of horizontal structure Roof characteristics Floor structure Type of foundation
6.	Damage classification Posting classification Geotechnical conditions of the ground
7.	Numbers of the photographs taken

Fig 3.5 : Overview of the information required in a form

The collection of data on the kind of materials used in the severely damaged buildings should be considered in order to make an assessment of the kind and amount of demolition waste materials, which can be used in site clearing and recycling plans. Regarding this subject, a suitable strategy should be developed. There is little problem in extending the Detailed Evaluation form in order to get information on the kind and amount of debris that can be expected. Information about the structural system gives an indication. It is, however, during the Rapid Evaluation that the most severely damaged buildings should be pointed out for demolition. Enlarging the Rapid Evaluation form to gain information about the debris is therefore necessary but should stay at a minimum so as not to obstruct the fast assessment work.

Another interesting element, brought up by Gavarini & Angeletti [3], is the environmental risk. An evaluation form should consider the safety of buildings or areas due to heavily damaged buildings in the neighbourhood. It is very important that those dangerous buildings are demolished as soon as possible.

Survey manual

A damage survey manual is a collection of assessment guidelines developed together with the survey form. The guidelines should be developed so that filling in the form becomes an easy and nearly automatic operation. The aim of the manual should be to make the survey repeatable and checkable and to obtain objective and reliable information. Developing a universal manual might be impossible because it will be influenced by different local ethnic, political and economic conditions. However, a manual should contain guidelines about three main subjects:

- The manual should give guidelines about how to make the damage assessment: where to look, and how to fill in the form.
- The manual should provide a classification of damage (3.4) in different levels in accordance with some qualitative or/and quantitative parameters.
- The manual should give information about the posting system (3.5) being used, and give practical guidelines about how to post the building.

Examples

The most complete manual we came across is the ATC manual [2], which is a compact overview of the assessment method presented in the ATC report. This manual, illustrated with many examples and photographs, is prepared for use by the surveyor in the field. It also contains a good damage survey form as presented in Figure 3.6. Both are discussed more in detail in 3.6 on page 75. Other manuals we came across in the literature were all summarized in tables. For other examples of survey forms and the accordant guidelines we refer to 3.6.

Block_____ Parcel No._____

ATC-20 Detailed Evaluation Safety Assessment Form

BUILDING DESCRIPTION :
Name : _____
_____
Address : _____
_____

No. of Stories :
Basement : Yes ❏ No ❏ Unknown ❏
Approximate Age : _____ Years
Approximate Area : _____ Square feet

Structural System :
Wood frame ❏ Unreinforced Masonry ❏
Reinforced Masonry ❏ Tilt-up ❏
Concrete frame ❏ Concrete Shear Wall ❏
Steel frame ❏ Other _____

Primary Occupancy :
Dwelling ❏ Other Residential ❏ Commercial ❏
Office ❏ Industrial ❏ Public Assembly ❏
School ❏ Government ❏ Emer. Serv. ❏
Historic ❏ Other _____

OVERALL RATING : (Check One)
INSPECTED (Green)   ❏
LIMITED ENTRY (Yellow) ❏
UNSAFE (red)   ❏

INSPECTOR :
Inspector ID _____
Affiliation _____

INSPECTION DATE :
Mo/day/year _____
Time _____am pm

Instructions : Complete building evaluation and checklist on next page and then summarize results below.

Posting :	Existing	Recommended	
None	❏		Posted at this Assessment
Inspected (Green)	❏	❏	❏ Yes ❏ No
Limited Entry (Yellow)	❏	❏	Existing posting by :
Unsafe (Red)	❏	❏	_____

Recommendations :
❏ No furhter action required
❏ Engineering Evaluation required (circle one) Structural Geotechnical Other _____
❏ Barricades needed in the following areas _____
_____

❏ Other *(falling hazard removal, shoring/bracing required, etc.)* :_____
_____

Comments *(Why posted Unsafe. etc.)* : _____
_____
_____
_____

Fig. 3.6 : ATC-20 Detailed Evaluation Safety Assessment Form [1]

ATC-20 Detailed Evaluation Safety Assesment Form (Continued)

Instructions : Examine the building to determine if any hazardous conditions exist. A "yes" answer in categories 1, 2 or 4 is grounds for posting building UNSAFE. If condition is suspected to be unsafe and more review is needed, check appropriate Unknown box(es) and post LIMITED ENTRY. A "yes" answer in category 3 requires posting and/or barricading to indicate AREA UNSAFE. Explain "Yes", "Unknown" findings and extent of damage under "Comments".

	Condition	Yes	No	Unknown	Comments
				Hazardous Condition Exists	
1.	*Structure Hazardous Overall*				
	Collapse/partial collapse	❏	❏	❏	_____
	Building or story leaning	❏	❏	❏	_____
	Other_____	❏	❏	❏	_____
	_____	❏	❏	❏	_____
2.	*Hazardous Structural Elements*				
	Foundations	❏	❏	❏	_____
	Roof/floors (vertical loads)	❏	❏	❏	_____
	Columns/pilasters/corbels	❏	❏	❏	_____
	Diaphragms/horizontal bracing	❏	❏	❏	_____
	Walls/vertical bracing	❏	❏	❏	_____
	Moment frames	❏	❏	❏	_____
	Precast connections	❏	❏	❏	_____
	Other_____	❏	❏	❏	_____
	_____	❏	❏	❏	_____
3.	*Nonstructural hazards*				
	Parapets/ornamentation	❏	❏	❏	_____
	Cladding/glazing	❏	❏	❏	_____
	Ceilings/light fixtures	❏	❏	❏	_____
	Interior walls/partitions	❏	❏	❏	_____
	Elevators	❏	❏	❏	_____
	Stairs/exits	❏	❏	❏	_____
	Electric/gas	❏	❏	❏	_____
	Other_____	❏	❏	❏	_____
	_____	❏	❏	❏	_____
4.	*Geotechnical Hazards*				
	Slope failure/debris	❏	❏	❏	_____
	Ground movement, fissures	❏	❏	❏	_____
	Other_____	❏	❏	❏	_____
	_____	❏	❏	❏	_____

SKETCH : . . . .

Fig. 3.6 : ATC-20 Detailed Evaluation Safety Assessment Form [1]

3.4 Classification of damaged structures

General

Once the degree of damage to structural elements has been assessed locally, attention must be paid to the effect on the structure as a whole. At this stage the surveyed structure should be classified according to a proposed damage classification system. This classification constitutes a major index for later decisions concerning the posting, repair, demolition and reconstruction phase, and will have an important influence on the amount of compensation which will be disbursed later.

Classification information together with detailed data about the structural systems, size of buildings, function and owners, provides the overall data showing the scale of the problem and the specific measures, with priorities, to be undertaken to reduce the consequences of earthquakes. The information can also be used for statistical analyses which can provide a guide in the development of new seismic design criteria.

Many different damage classification systems can be found in the literature, varying in the kind of classification guidelines and the number of classification levels. Mostly they are also written in terms of the aim they were developed for.

There are two kinds of classification guidelines based on qualitative and quantitative judgement criteria respectively. A classification system based on quantitative judgement criteria defines the different damage levels, by a precise range of values that some objective and measurable quantities take on. Such quantities are, for example, type and width of cracks, settlement and tilt angle. This identification of the damage level should be objective, checkable and repeatable, and should not be influenced by personal judgement. On the other hand, measurements and thus, measuring instruments, are necessary. A classification based on qualitative criteria does not use measurements to evaluate damage, but is based on a visual assessessment only. This makes the classification less objective but makes a quicker evaluation possible. Most of the classification systems, however, use a combination of quantitative parameters and a qualitative judgement.

After a disaster some buildings will remain undamaged, others will be partially or totally collapsed and a third group will contain a wide range of damage conditions. On the one hand we need a minimum of three

classification levels, whilst on the other hand, the damage assessment
has to be based on a visual examination. So an appropriate classification
system cannot have too many damage levels, otherwise the range of
decision would become too small.

The Rapid Evaluation, based on a quick outside examination of the
building uses the minimum number of levels: safe (undamaged or
slightly damaged), unsafe (very severely damaged or collapsed), and
doubtful (medium damage) buildings. The Detailed Evaluation is a more
thorough visual evaluation, and will give more information about the
damage condition of the structure. This makes it possible to use a finer
classification system with more damage levels.

Examples

Most damage classification systems for the second evaluation level, that
we came across in the literature, use five or six damage levels. This
paragraph summarizes some examples. Some of them are described in
more detail in 3.6 .

1. A good example of a classification system for reinforced concrete
buildings is being prepared by the Japanese Ministry of Construction [5]
as mentioned in 2.10. It classifies the damaged buildings in five damage
degrees (Fig. 3.7), following quantitative and qualitative guidelines. A
similar one for wooden structures is presented in Fig. 3.17.

2. Another methodology using six damage levels is proposed by
Petrovski [4] (Fig. 3.8). A description is given in 3.6.

1.	Undamaged structures.
2.	Slightly damaged structures, with slight nonstructural damage and negligible damage to the structural system.
3.	Buildings with damaged structural system.
4.	Buildings with considerably damaged structural system.
5.	Buildings with heavily damaged structural system.
6.	Partially or completely failed buildings.

Fig. 3.8 : Six damage levels proposed by Petrovski [4]

3. That five damage levels seem to be a good practical choice was
proven during the damage assessment after the earthquake of El Asnam
[10] (Algeria 1980). After the disaster, the damage of all the buildings

Damage rate	Characteristics of damages	Judgement by the settlement S(m) of the building	Judgement by the tilt angle q (radian) of the building	Judgement by the damage rate of the building
Slight	Some parts of non-bearing walls have cracks. Some parts of girders, columns and walls have visible cracks.			Crack widths are so small that it is difficult to recognise them with the naked eye (crack width is under 0.2 mm).
Small	Non-bearing walls have cracks. Girders, columns and walls have visible cracks, and some members partially have substantial cracks.	$S \leq 0.2$	$q \leq 0.01$	Clearly visible cracks can be seen with the naked eye (crack width is between 0.2 and 1 mm).
Medium	Many columns and walls show remarkable shearing or bending cracks, and many members have partial shearing damage and buckling in axial reinforcing bars.	$0.2 < S \leq 1.0$	$0.01 < q \leq 0.03$	Comparatively large cracks form, but fallen off concrete is not substantial (crack width is between 1 and 2 mm).
Severe	Main bearing walls are broken, and the axial reinforcement of columns is exposed by buckling, while the structure is not collapsed.	$S > 1.0$	$0.03 < q \leq 0.06$	The number of large cracks (more than 2 mm) and fallen off pieces of concrete is severe, reinforcement is exposed.
Collapse	Many columns and walls are fallen off and have failed due to buckling of axial reinforcing bars. The building is totally or partially collapsed.		$q > 0.06$	Reinforcement bars are buckled and internal concrete of reinforcement is crushed. Deformation along the vertical direction of the columns (or bearing walls) can be seen at a glance. The settlements and tilts of building can be a feature of collapse. In some cases, reinforcement is broken.

Fig. 3.7 : Damage classification for reinforced concrete structures [5]

in the city was evaluated by a hundred engineers of the C.T.C. (Algerian Organisation of Technical Control of Construction). The damage was noted on special forms and the different damage conditions were finally divided in five categories (Fig. 3.9).

1	No damage	
2	Slight damage	non-structural damage
3	Moderate damage	important non-structural damage and slight damaged structure
4	Severe damage	severe non-structural damage and important structural damage
5	Collapse	

Fig. 3.9 : Damage classification used in El Asnam [10]

4. Another method used by the Californian Office for Emergency Services [1] rates the damage to certain structural elements on a seven level scale, corresponding to the percentage of damage estimated (Fig. 3.10). This method gives no definitions of damage states or guidelines to distinguish between the different ratings on the damage scale, particularly the ratings 2, 3 and 4, which are all "moderate" damage. It relies heavily on the engineering knowledge and the judgement of the inspector, and is therefore not a very good choice.

The structural engineering association, SEAONC, developed guidelines to be used with this rating scale. They give examples of damage that would correspond to the various damage ratings.

0	None	0%
1	Slight	1-10%
2,3,4	Moderate	11-40%
5	Severe	41-60%
6	Total	over 60%

Fig. 3.10 : COES classification system [1]

5. Other classification systems do not classify the building as a whole, but classify some structural and non-structural elements of the building. This is a more thorough evaluation, but will consume a lot more time. In our opinion these classification systems would not be useful in the relief phase, but belong instead to the more detailed Engineering Evaluation level.

An excellent example is given by the CEB-report [9] from 1983 regarding the assessment of concrete structures. It distinguishes five levels of damage of reinforced concrete elements due to earthquakes. They correspond implicitly to levels of available safety margins after damage, as well as to certain classes of repair and/or strengthening measures to be taken. In Figure 3.11 the damage classification of vertical bearing elements is presented. For other structural elements (such as beams and slabs), similar levels of damage could be adopted.

A-level	Isolated flexural cracks presenting widths less than about 1 to 2 mm, provided that a simple computation has proved that these cracks are not due to steel-section inadequacy, but rather to local deficiencies (e.g. construction joints, inadequate anchorages, restraint due to partition walls, slight shocks, etc.).
B-level	Many large flexural cracks or isolated diagonal shear cracks (presenting widths less than about 0.5 mm), provided that permanent deflections have not been noticed at all.
C-level	Bidiagonal shear cracks and/or intensive local spalling of concrete due to shear and compression, provided that no appreciable residual displacements have been observed. (Cracking within beam-to-column joints is considered as C-level damage).
D-level	Core concrete failed, steel bars buckled (the building-element is discontinued but not collapsed), provided that little residual deflections (both vertical and horizontal) have been observed. (Severe disintegration of beam-to-column joints is considered as D-level damage).
E-level	Partial collapse of vertical elements.
If the conditions set forth for residual deflections are not fulfilled, the corresponding damage level is increased by one level.	

Fig. 3.11 : CEB damage levels for vertical bearing elements [9]

6. Another example that does not classify the building as a whole is proposed by Braga, Liberatore & Dolce [3], and was used for the damage assessment after the Southern Italy earthquake of 1980. It classifies the damage of six structural and non-structural elements of the building in six different levels according to the value of some objective and measurable quantities (see Fig. 3.19 on page 30). These elements are the vertical structure, the floors, the roof, the outside walls, the partition walls and the stairs. Additionally, a distinction is made between elements of different types of structures.

3.5 Posting of the building

General

We have already shown that a major task for the building survey team is to inform the civilians about the condition of their dwellings, the surrounding buildings, the lifeline facilities (gas, water, electricity), and the safety of the area. It can reassure the people about the safety of their houses and can avoid further victims by uncontrolled reuse of damaged buildings. An important tool for doing this, is posting the buildings after the assessment and barricading dangerous areas.

Posting buildings

Most of the publications choose a three level usability posting system: inspected or safe, unsafe and limited entry. Examples of these are the ATC-20 [1] and the Petrovski [4] report. This sytem is found the best choice and the most straightforward to use. A system that only uses the two posting possibilities, safe and unsafe, is found difficult or even impossible to use, because of the lack of structural engineers and the immense number of buildings to be quickly examined. When the assessments do not use sophisticated assessment techniques there will always be "grey-area" buildings. So an intermediate posting level, designated Limited Entry, is necessary for buildings that need a closer examination. On the other hand, using more posting levels can be confusing, to the inspectors as well as to the public.

Inspected / Safe	Green	No apparent hazards found, although repairs may be required. Original lateral load capacity not significantly decreased. No restriction on use or occupancy.
Limited Entry	Yellow	Dangerous condition believed to be present. Entry by owner permitted only for emergency purposes and only at own risk. No usage on continuous basis. Entry by public not permitted. Possible major aftershocks hazard. (Building needs further examination).
Unsafe	Red	Extreme hazard, may collapse. Imminent danger of collapse from an aftershock. Unsafe for occupancy or entry, except by authorities. (Building should be demolished as soon as possible).

Fig 3.12 : The most common building posting system [1]

Posting buildings can be done by different methods, using placards or painted marks. One element, however, that all the methods have in common is the use of colours. Depending on the damage level, the buildings are posted in a different colour. The three main colours are Green, Red and Yellow, which are used worldwide for entrance posting (e.g. in the traffic). The most common building posting system [1] is presented in Figure 3.12.

The "Uniform Mitigation Plan" (International Conference of Building Officials, 1979) [1] contains an interesting, extended posting procedure. It has five possible usability postings, with five different corresponding colours (Fig. 3.13). The difference with the procedure before, is that habitable dwellings requiring some repairs are not posted green, but yellow. This gives extra information for the owner of the building. The blue colour gives extra information about the usability of all the lifelines. This is an interesting consideration because the inspection and repair of lifelines, especially the gas connection, often suffer major delays (see 2.7). So, to make a distinction between safe buildings with repaired and with non-repaired lifelines, a supplementary colour should be considered.

Red	Keep out, uninhabitable.
Gold	Limited Entry permitted, but no occupancy. Owner may enter at own risk to remove property.
Yellow	Habitable, repairs required.
Green	Safe for occupancy.
Blue	Approved for connection of water, electricity and/or gas.

Fig 3.13 : The five posting classes of the Uniform Mitigation Plan [1]

Posting guidelines should be part of the survey manual together with the damage classification guidelines. To make the building posting procedure simpler and quicker it is important that there is a link between the posting decision and the damage classification. A good example is presented by Petrovski [4]. The buildings should be posted according to the usability and the damage level (Fig. 3.14).

Colour	No.	Usability	Damage category	Mark
Green	I	Usable	Undamaged buildings	One green line
		Usable	Slightly damaged building	Two green lines
Yellow	II	Temporarily unusable	Buildings with damaged structural system	One yellow line
		Temporarily unusable	Buildings with considerably damaged structural system	Two yellow lines
Red	III	Unusable	Buildings with heavily damaged structural system	One red line
		Unusable	Partially or totally failed buildings	Two red lines

Fig. 3.14 : Posting connected with damage classification (Petrovski [4])

Posting the building not only gives information about the building's safety but it also indicates that the specific building has already been inspected. Therefore an appropriate method of posting will, beside usability information, also gives information about the inspection date, the inspection team and the location of the building. There are many ways to do the posting. We can give two different examples:

a) The first method proposed in the ATC-20 report [1] posts the buildings (Inspected, Limited Entry or Unsafe) with placards. The placards are in the colour corresponding to the safety level and carry additional information about the inspection date, the inspection team and the building location. These placards give distinct information to the public, because they use significant text giving clear warnings. A major disadvantage of the use of placards is that they can easily be removed.

b) Thus, painted marks on buildings can provide a better solution. This method uses a specific code marked on the building in the colour corresponding to the safety level of the building. The code (Fig.3.15) proposed by Petrovski [4] comprises the number of the colour (I, II, or III), the identification number of the Working Group (105), the identification number of the inspected building corresponding to the assessment form (406) and a number of lines (1 or 2) corresponding to the damage level (Fig. 3.14). The proposed code has a better link with the assessment forms. This is very important, because later decisions about demolition, repair and financial relief are made using these forms. A disadvantage of the proposed mark is that it is not accompanied with a clear warning sign, but this can easily be accomplished.

I.105/406

Fig. 3.15 : Posting code proposed by Petrovski [4]

Whatever the method, the actual posting should be done by placing the appropriate placard or mark in a clearly visible place near the main entrance, and by placing additional ones at all other entrances for buildings posted Unsafe or Limited entry.

Posting unsafe areas

In addition to the usability posting of buildings, there is also a need to designate certain areas as unsafe. These may either be inside or outside the building. These areas must be cordoned off with yellow "Do Not Cross Line" tape (only used for this purpose) or otherwise barricaded to prevent entry. For example, if a falling hazard is observed, such as a badly damaged parapet, the area within potential striking distance must be barricaded. Other examples are toxic chemical spills, damaged canopies, damaged elevators, fallen powerlines, etc.

It is very important to consider that badly damaged buildings can be a danger for a whole area including other buildings. So safe buildings can be part of barricaded areas not allowing entry. Therefore, it is important for the rehabilitation of the neighbourhood that hazardous elements and collapsing buildings are demolished as fast as possible.

3.6 Damage assessment of buildings

General

Since the technical expertise of the persons entrusted with the investigation of the damage will inevitably cover a very broad range, inspections should be carried out using a unified method or, at least, a standard approach to the question of assessing the extent of damage to structures.

This method should be summarized in a manual which will guide the surveying engineer in the field. The degree and depth of investigation will vary from one structure to another. Preliminary visual inspection, or Rapid Evaluation, should point out those structures which are safe for reuse as well as those which clearly cannot be reconstructed, but need only proper and efficient demolition and dismantling. Their identification will make it possible for the specialists to concentrate on the structures that need a deeper investigation. That is where the Detailed Evaluation begins. Here too, as the number of structures is large, it is desirable that the investigation should take the form of a questionnaire, so that the requisite information can be processed in a standardized manner and within the shortest time possible. After the assessment, the surveyor will be able to classify the structure regarding the degree of damage and finally post the structure, in accordance with the classification and posting guidelines, which are also part of the manual.

When serious damage is suspected and not enough of the structure is viewable to permit a reliable visual evaluation, the building needs an Engineering Evaluation.

Examples

Below are some interesting assessment procedures, belonging to the second evaluation level, described. All the basic subjects previously mentioned will be reconsidered.

1. Applied Technology Council [1]

The ATC assessment method was specially developed for the safety and usability evaluation of earthquake damaged buildings. In this regard it is probably the best assessment method to date. The main character of the method is the three level structure (Rapid, Detailed and Engineering Evaluation), already discussed in 3.2. Here we will take a closer look at the second evaluation level.

For most buildings the principal objective of the Detailed Evaluation is to establish whether there is a possibility of either structural collapse or falling hazards. These are the two major life-safety concerns when evaluating a damaged structure. A fundamental assumption is to consider that the structure must be capable of withstanding at least a repetition of the event. This is, however, a minimum requirement.

The structural system of a building can be thought of as consisting of two systems: the vertical-load-bearing system and the horizontal-load-bearing system. For a structure to be considered safe, both systems must be functional and cannot be seriously degraded. For a typical building this can be examined by inspecting the architectural walls, on both the interior and exterior of the building, to verify that destructive storey drifts have not taken place. This information can be supplemented by observations of the structural system, wherever it is exposed, such as in basements, stairwell, etc.

If it were not for the fact that architectural elements, such as walls and ceilings, hide the structural system of most buildings, it would be a relatively easy matter to make post event inspections and determine the extent of any structural damage. Unfortunately this is not the case in most buildings. When serious damage is suspected and not enough of the structural is viewable to permit a reliable evaluation the building should be posted Limited Entry or Unsafe. The owner will then have to arrange an Engineering Evaluation.

The list below (Fig. 3.16) presents the general guidelines for rating damaged buildings. These are factors of concern common to many different types of buildings. The manual [2] gives, in addition, more structure- or situation-specific guidelines covering where to look for damage, how to rate its usability, and what posting category to use; respectively for wood frame, masonry, tilt-up, concrete, and steel frame structures. Geotechnical hazards and non-structural hazards are covered in two additional chapters.

In accordance with these guidelines the building inspector should answer the question "Does this specific hazardous condition exist ?" for the different factors of concern, with three possibilities: Yes, No and Unknown. This answers should be noted on the accompanying form (Fig. 3.6).

1. Overall damage: This is the best indicator of a severe damaged structural system. If a Rapid Evaluation has not been done, this evaluation is begun by examining the entire building, inside and out, for the unsafe conditions here below:	
- Collapse or partial collapse	UNSAFE
- Building or individual story noticeably leaning	UNSAFE
- Fractured foundations	UNSAFE
2. Vertical load system: Failure of the vertical load bearing system either globally or locally is generally considered grounds for posting the entire structure unsafe.	
- Columns noticeably out of plumb	UNSAFE
- Buckled or failed columns	UNSAFE
- Roof or floor framing separation from vertical supports	UNSAFE
- Bearing wall, pilaster, or corbel cracking that jeopardizes vertical support	UNSAFE
- Other failure of vertical load carrying system	UNSAFE
3. Lateral load system: First identify and then inspect the system. To be permitted to remain in use, it must have a functioning lateral load system.	
- Broken, leaning, or seriously degraded moment frames	UNSAFE
- Severely cracked shear walls	UNSAFE
- Broken or buckled vertical braces	UNSAFE
- Broken or seriously damaged diaphragms or horizontal bracing	UNSAFE
- Other failure of lateral load carrying system	UNSAFE
4. P-Delta effects: For tall frame structures, particullarly high-rise buildings, any residual storey drift is generally quite serious. The weight of the portion of the structure above the deflected storey results in an additional moment on the columns due to the P-delta effect.	
- Multistory frame buildings with residual storey drift	UNSAFE
5. Degradation of the structural system: It is important to determine that the entire system has not been so degraded that its strenght and stiffness have been reduced to unsafe levels. This is a particular concern for concrete and masonry structural systems.	
- Seriously degraded structural system	UNSAFE
6. Falling hazards: The post-event hazard of damaged items, as parapets, cladding, ornaments, signs, ceilings and lightfixtures, may fall as a result of static forces or an aftershock. Areas within striking distance must be placed off limits and barricaded.	AREA
- Falling hazards present	UNSAFE
7. Slope or Foundation Distress: Examine the ground in the immediate area of the building for evidence of mass ground displacements. These vertical or horizontal ground displacements may fracture foundations and cause severe structural distress to building superstructures. Often, geotechnical hazards cover areas larger than a single structure.	
- Base of building pulled apart or differentially settled, with fractured foundations, walls, floors, or roof	UNSAFE
- Building in zone of faulting or suspected major sloop movement	UNSAFE
- Building in danger of being impacted by sliding or falling landslide debris from upslope	UNSAFE
8. Other hazards: Unsafe conditions such as friable asbestos release, broken fuel line, chemical spill, downed power line.	AREA UNSAFE

Fig. 3.16 : ATC Inspection list for damaged buildings [1]

The assessment results in posting the building with placards (see 3.5). Safe structures are posted inspected (green colour), the unsafe structures are posted unsafe (red) and the potentially dangerous structures posted limited entry (yellow). The damage inspection guidelines require the use of judgement. Under some circumstances the use of a less restrictive posting or other action may be appropriate.

2. Japanese Ministry of Construction, 1986 [5]

This standard contains five damage degrees: slight damage, small damage, medium damage, severe damage and collapse. The evaluation is based on the judgement of quantitative parameters (settlement of the building, tilt angle of the building and a damage rating as a function of crack openings) together with some visual inspection of the characteristics of the damage. The main procedures for the evaluation of the damage are the measurement by eye and deformation. A summary of the evaluation and classification of the damage rate for reinforced concrete stuctures has already been presented (Fig. 3.7). Figure 3.17 shows a similar table for wooden structures.

3. Jakim Petrovski, Skopje, Macedonia [4].

The classification of the damage level and the estimation of the usability of buildings within earthquake affected areas is carried out on the basis of the detailed damage description, according to six categories (Fig. 3.18). The first two categories comprise undamaged buildings, as well as buildings of which the original seismic stability has not been decreased. Buildings belonging to these categories are allowed to be used immediately after the earthquake, i.e. as soon as the damaged members, such as chimneys, attics and gable walls have been removed or repaired.

The buildings belonging to the next two categories have decreased seismic stability. These buildings should be repaired by strengthening of the structural elements. The need for supporting and protection of both the building and its surroundings should be considered. It is not permitted to use the building.

The last two categories are buildings subjected to sudden collapse. The building's surroundings should be protected and access should not be allowed. The final decision for demolition of the building should be made on the basis of a study of technical possibilities and economical justification for repair of the structure.

Damage	Damage Characteristics	Judgement of members				
		Foundation	Floor frame	Framework	Wall in framework	Roof
Slight	Slight slipping off at the joint of girder and column. No damage at the framework of columns and girders. No settlement.	2 to 5 cracks over 0.3 mm in width and less than 200 mm in lenght.	Slight slip off between the floor frame and the wall.	Slight slip off at the joint between girder and column.	Slight slip off in the wall.	Slip off at a part of ridge roof tiles.
Small	Damage at some joints of girders and columns. Girders and columns show slight deflection. Almost no residual deformation in walls. No differential settlement.	Slightly larger than mentioned above.	Gaps at the joint of floor panels.	Damage of compressive strain inclined to the grain at the joint of girder and column.	Slight cracks on the wall.	Slip off and fallen out at the ridge roof tiles.
Medium	Damage at the framework of columns and girders. There is slipping off at the joint of girder and column, and residual deformation in every framework. Differential settlement.	Different settlement, local damage, removal and fell off of finishing mortar.	Some ruggedness on the floor.	Check and shake and slip off can be seen at the joint of girder and column.	Residual deformations on nearly all walls in the framework.	Slip off both at the roof tiles and at the ridge roof tiles.
Severe	The framework of columns, girders and others is broken. The building does not collapse, but forms large residual deformation. Braces in walls are broken, and walls are removed off from the columns and sills.	Substantial cracks and breaking.	Substantial ruggedness and breaking can be seen on the floor.	Partial defect in cross section of girders and columns. Breaking and residual deformation at girders, columns and ceiling.	Substantial residual deformations on nearly all walls in the framework.	Ruggedness can be seen at the roof, and the roof truss is partially damaged.
Collapse	All columns broken, and fallen off from the sill. A residual deformation unable to recover. First story completely collapsed.	Movement, overturning occured. Foundations can no longer support the upper structure.	Substantial ruggedness at all floors. All joists and sleepers are fallen out.	All columns are snapped and the sill is removed from the foundation.	Large residual deformation and collapse.	Substantial damage at the roof truss, and at whole roofing materials.

Fig. 3.17 : Damage classification for wooden structures [5]

Damage Category	Damage Description
Undamaged structures	Without visible damage to the structural members. Fine cracks in the wall and ceiling mortar.
Slightly damaged structures	Structures with slight nonstructural damage and negligible damage to structural system. Cracks to the wall and ceiling mortar. Falling off of large patches of mortar from wall and ceiling surfaces. Considerable cracks and damage, or partial failure of chimneys, attics and gable walls. Disturbance, partial sliding, sliding and falling down of roof covering. Cracks in structural members.
Buildings with damaged structural system	Diagonal or other cracks to structural walls. Diagonal cracks to walls between windows and similar structural elements. Large cracks to reinforced concrete structural members: columns, beams, RC walls. Heavily damaged, partially failed or failed chimneys, attics or gable walls. Disturbance, sliding and falling down of roof covering.
Buildings with considerably damaged structural system	Large cracks with or without disattachment of walls with crushing of materials. Large cracks with crushed material of walls between windows and similar elements of structural walls. Larger cracks with crushing of materials of RC structural elements: columns, beams and RC walls. Slight dislocation of structural elements and the whole building.
Buildings with heavily damaged structural system	Structural members are heavily damaged and dislocated. Connections of structural elements are heavily damaged. A large number of crushed structural elements. Considerable dislocations of the whole building. Significant deleveling of roof structure.
Partially or completely failed buildings	Considerable crushed structural members. Partially failed structural elements. Considerably dislocated, partially or completely failed buildings.

Fig. 3.18 : Damage classification by Petrovski [4]

The adopted methodology requires certain general physical data according to a given form, which should be completed in situ. The form comprises the necessary information to be used for statistical data processing on computer. Therefore, a code system was developed and is presented in the accompanying manual.

The posting is done (page 73) by marking the inspected building with colours according to the damage level category and the usability (Fig. 3.14 and 3.15).

4. Braga, Liberatore & Doce, Italy, 1984 [3]

The presented form and manual are derived from those used by the Italian Army during the damage survey after the Southern Italy earthquake of November 23, 1980. The form has got some modifications, aiming at a more accurate description of the structure, which do not complicate the survey operations. According to the authors, a team of two people, after one day of training, can survey 5 to 10 buildings per day, depending on the size of the buildings.

The methodology examines the level of damage of six types of structural and non-structural elements: vertical structure, horizontal structure, roof, outside walls, partitions and stairs, and it classifies these elements into six damage levels. These damage levels are determined on the basis of objective parameters such as the gravity, location and diffusion of cracks and crushing. The proposed manual is a table that summarizes the guidelines for identifying the damage level of the various elements. The manual is presented here together with the survey form in Figure 3.19. The form requires four groups of information concerning:

- The identification of the building and the survey team, and the composition of the building.
- The use of the building.
- The structural typology of the building, defined by the cross-combination between the horizontal and vertical structure. A distinction is made between buildings with resistant elements in both directions or in one direction only. The reinforced concrete buildings are divided into three typologies: shear walls, frame without and with soft storey.
- The damage classification for six types of structural and non-structural elements.

5. Californian Office for Emergency Services, 1988 [1]

Refering to the ATC report the OES document "Safety assessment plan for volunteer engineers" contains damage assessment report forms for eleven specific civil engineering structures, including buildings, airports, bridges, dams, etc. The damage is rated on a scale of 0 to 6, and the posting system is the same as proposed by ATC.

DATE ..../..../....   TEAM ..../..../...   NUMBER ..../..../....

LOCATION OF THE BUILDING

ADMIN. _____   DISTRICT ____
UNIT

STREET _____   No. _____

BLOCK _____   BUILDING _____   CARTOGRAPHIC REF. _____

COMMON ☐ 1   ATTACHED _____   CADASTRAL REF. SHEET _____
FRONTS ☐ 2   BUILDINGS _____   MAP _____
     ☐ 3

METRICAL DATA

AVER. AREA (SM) _____   AVER. HEIGHT (M) _____   VOLUME (CM) _____
STORIES ____ . ___   CELLAR __ . __   ATTIC __

PURPOSE OF THE BUILDING

DWELLING   No. OF FLATS _____   No. OF STAIRS ____   No. OF ROOMS _____

CONDUCTIVE   HANDICRAFT _____   TRADE _____   TOURISM ____   OTHER _____
ACTIVITIES

STRUCTURAL CHARACTERISTICS

VERTICAL STRUCTURE			HORIZ. STRUCT.		ROOF		AGE	
DIRECTION	1	2	VAULTS	☐	WOOD AND TILES	☐	YEAR _____ OR	
WALLS FIELD STONE	☐	☐					BEFORE 1900	☐
HEWN STONE	☐	☐	WOODEN	☐			1901-1943	☐
BRICK MASONRY	☐	☐	FLOORS		R/C	☐	1944-1962	☐
R/C	☐	☐					1963-1971	☐
			STEEL	☐	OTHER	☐	1971-1981	☐
FRAMES WITH SOFT STORY	☐	☐	FLOORS				1982-AFTER	☐
WITHOUT SOFT STORY	☐	☐			COMPOSITE	☐	UNKNOWN	☐
			R/C FLOORS	☐				

DAMAGE

DAMAGE	1	2	3	4	5	6	
VERTICAL STRUCTURE	☐	☐	☐	☐	☐	☐	1. NO DAMAGE
FLOORS	☐	☐	☐	☐	☐	☐	2. SLIGHT
ROOF	☐	☐	☐	☐	☐	☐	3. CONSIDERABLE
OUTSIDE WALLS	☐	☐	☐	☐	☐	☐	4. VERY SERIOUS
PARTITION WALLS	☐	☐	☐	☐	☐	☐	5. PARTIALLY COLLAPSED
STAIRS	☐	☐	☐	☐	☐	☐	6. COLLAPSED

Fig. 3.19 : Italian damage survey form and accompanying manual [3]

con. element	damage level 2 Slight	3 Considerable	4 Very serious	5 Partial collapse	Abbreviations
R/C beams and columns	C ≤ 1 mm	C > 3 mm CS partial OP ≤ 40 mm	C > 5 mm CS marked SL DU OP > 40 mm	PC	
bearing masonry walls	C ≤ 1 mm	C(1-2) ≤ 10 mm or C(5) > 10 mm	C(2) ≤ 10 mm or C(1) > 10 mm OP > 50 mm DJ from HS	PC	B = Beams BS = Beam supports BU = Reinforcement Buckling C = Cracks CC = Cross Cracks CS = Crushing in R/C cors DC = Diagonal Cracks DJ = Disjunctions
masonry vaults	C ≤ 1 mm in NTBV	C > 1 mm in NTBV	SH of KS C and CS in P	PC	DJ-VS = Disjunctions from vertical structure DSC = Disconnections F = Fail HCB = Hollow Clay blocks HS = Horizontal Structure
wooden floors	LC many CC some	DJ - VS of B plus suburden	SO of B PC	as 4 but MM	KS = Keystones LC = Longitudinal Cracks MM = More Marked NTBV = Vaults without tie beams
Iron floors	LC many CC some	LC large CC distribution DJ-VS of B	CS of BS CC distribution DJ-VS of B F of HCB	PC	OP = Interstory out of plumb P = Piers PC = Partial collapse PT = Partitions
R/C floors and roofs	LC many CC some DJ of HCS some	LC large CC many CS at BS DJ of HCB many	CC large CS evident DJ-VS CS of PT	PC	R/C = Reinforced concrete RS = Roof Scaffolding SD = slight Disconnecting SH = Shiftting SL = Reinforced slippage
wooden roofs	SH of B DSC in TC	SH of B DSC in RS F of TC partial	SO of B PC	as 4 but MM	SO = Slipping off SP = Spelling of cover TBV = Vaults with tie beams TC = Tile Covering
external walls and partition	DC ≤ 1 mm DJ-VS ≤ 1 mm	DC ≤ 10 mm DJ-VS ≤ 10 mm local coil	as 3 but MM	PC	(S) = Single (1) = Distributed on at least 1/3 of surface (2) similarly (2/3) (1-2) similarly (1/3 to 2/3)
stairs	SD	DSC evident DC and SP in R/C B	PC	as 4 but MM	

Fig. 3.19 : Italian damage survey form and accompanying manual [3]

STATE OF CALIFORNIA

OES/PEO

Assessment

BUILDING/STRUCTURE Report No _____

Facility Name _____
Address _____
Co-City-Vic _____
Mo/Day/Yr ___ / ___ / ___ Time ___
Type of Disaster _____

OES/PEO ID No's _____ _____
Other PEO Reports _____
No. Photos _____ No. Sketches _____
Ref. Dwgs. _____
Est. Damage $ _____

Facility Status

SAFETY INSTRUCTIONS : The possiblity of the presence of toxic gases in confined spaces or of fuel leaks should be recognized as potential hazards.

CAUTION : This report was made for the California State Office of Emergency Services. The prime purpose of the report is to advise of the condition of the facility for immediate continued use/occupancy. REINSPECTION OF THE FACILITY IS RECOMMENDED. AFTERSHOCKS MAY CAUSE DAMAGE WHICH REQUIRES REINSPECTION. The conclusions reached by engineers who re-examine the facility later should take precedence. The assessment team will not render further advice in the event of conflict of engineering recommendations.

A. PLACARD POSTING
  1. Existing : None ❏ 2. Recommended : Green ❏ 3. Posted at this assessment : Yes ❏
           Green ❏   posting   Gold ❏   No ❏
           Gold ❏   Red ❏
           Red ❏
  Posting authority : _____

B. RECOMMENDATIONS
  1. Shoring and bracing   2. Barricades needed   Yes ❏ No ❏
           Not apparently needed ❏   3. Monitor building movements Yes ❏ No ❏
           Needed to protect public ❏   4. Monitor weakened components Yes ❏ No ❏
  Needed to protect adjacent building ❏   5. Structural Report by CE or SE
                                    recommended   Yes ❏ No ❏

C. COMMENTS _____
_____
_____
_____
_____
_____

Fig. 3.20 : Form developed by the Californian Office for Emergency Services [1]

D. FACILITY DESCRIPTION

Assessment Report # _____

	Lateral support system (vertical)	Exterior walls	Horiz. diaphragm	Horizontal bracing (steel)	Interior bracing and/ or shear walls	Non-bearing partitions	Floors	Stairs	Other ( )
Concrete									
Precast concrete									
Reinforced masonry									
Unreinforced masonry									
Steel frame									
Metal deck									
Wood									
Other									
Unknown									

1. Number of stories _____
2. Number of basement levels _____
3. Pre-1934 design ?
   Yes ⊐   No ⊐   Unk ⊐
4. Primary occupancy :

Government ⊐	Hotel/motel ⊐
Hospital ⊐	Commercial ⊐
School ⊐	Industrial/mfg ⊐
Office bldg. ⊐	Residential ⊐
Other _____	

E. STRUCTURAL DAMAGE OBSERVATIONS (D.O.)

Damage scale :

0	1	2-3-4
None (0%)	Slight (1-10%)	Moderate (11-40%)
5	6	NA	NO
Severe	Total	Not	Not
(41-60%)	(Over 60%)	Applicable	Observed

		D.O.
1.	Exterior walls	_____
2.	Frame (general condition)	_____
3.	Frame members	_____
4.	Frame connections	_____
5.	Roof framing	_____
6.	Interior bearing/shear walls	_____
7.	Partitions (non-bearing)	_____
8.	Floor(s)	_____
9.	Stair(s)	_____
10.	Elevator	_____
11.	Glass	_____
12.	Mechanical equipments supports	_____
13.	Electrical eqipment supports	_____
14.	Other	_____
15.	Total % damaged	_____

F. SOIL OR GEOLOGICAL DAMAGE

	Appar. Hazard	No Appar. Hazard
1. Settlement	_____	_____
2. Liquefaction	_____	_____
3. Landslide	_____	_____
4. Faulting	_____	_____
5. Other	_____	_____

G. FALLING HAZARDS

	Appar. Hazard	No Appar. Hazard	Unknown
1. Parapet walls	_____	_____	_____
2. Ornamentation	_____	_____	_____
3. Chimney (s)	_____	_____	_____
4. Floor	_____	_____	_____
5. Roof structure	_____	_____	_____
6. Equipment	_____	_____	_____
7. Other	_____	_____	_____

H. ESTIMATE OF DAMAGE

1. Approx. bldg. area ( ) _____
2. Est. % of bldg. damaged _____
3. Est. damage valuation $ _____

I. REMARKS AND SKETCHES

Fig. 3.20 : Form developed by the Californian Office for Emergency Services [1]

The OES form (Fig. 3.20) for general buildings is considered as good. Structural elements are rated on a damage scale from 0 to 6, with 6 corresponding to over 60 % of damage (page 69). The form also contains a hazard evaluation of falling items as well as soil and geologic conditions. However, the OES forms rely heavily on the engineering knowledge and judgement of the inspector. There are no definitions of damage states or guidelines to distinguish between different ratings on the damage scale. In addition, there are no obvious correlations between several of the damage ratings and the safety posting.

In 1987, the Structural Engineers Association of Northern California Disaster Emergency Services Committee developed a draft document titled "Earthquake damaged building assessment criteria". It contains a brief but comprehensive set of guidelines to be used with the OES form They provide evaluation criteria for the principal structural elements as well as falling hazards and geotechnical conditions. The guidelines give examples of damage to these elements that would correspond to the various damage ratings. The manual also connects the damage ratings to the posting colours: red to damage states 6 and 5, yellow to 4 and 3, and green to 2, 1 and 0.

6. Gavarini and Angeletti, Italy, 1984 [3]

The authors present a totally different approach. Their inspection goal is the assessment of the risk due to aftershocks in periods following the major quake. A qualitative assessment procedure is chosen because a quantitative safety assessment, taking account of the many variables involved in such a very limited time and in emergency conditions, is not yet possible. The following successive steps seem necessary for attempting the final goal:

- Assumption of the maximum expected intensity of actions on the area under consideration (aftershocks).
- Assessment of environmental vulnerability in post-shock conditions. This is not just defined by life line failures, landslides and dam failures but also - what is peculiar to urban areas and old towns - by hazardous escape ways, dangerous close buildings, etc.
- Assessment of building vulnerability in post-shock conditions.
- Assessment of the type and the level of damage considering the main shock and the expected aftershocks.

A tentative approach for assessing the building's safety during the emergency is presented in terms of questions. The elements which provide inspection teams with guidelines for final judgement and posting are part of the survey form. This form is organized in the following way: the first column lists elements of hazards with associated keywords for criteria to be followed through the judgement process; the second column is to be filled by notes and observations; the third column shows the degree of importance of each element (high, medium, low) for final judgement; the fourth column will contain the answer to the question "Is the actual element hazardous ?" with five different possibilities:

- Yes, the hazard is high.
- Yes, the hazard is medium.
- Yes, the hazard is low.
- No, it is not.
- I do not know.

3.7 Essential facilities, infrastructure and lifelines

Essential facilities

After a disaster with many casualties and many people left homeless, services provided by police and fire departments, disaster relief organizations, hospitals and medical organisations will be badly needed. For maximum effectiveness, these emergency service organizations must operate from safe facilities. Those facilities most needed by a community following a disaster, called essential facilities, commonly include: hospitals, health care facilities, police and fire stations, jails and detention centers, communication centers, emergency operations centers, etc.

The best solution of course is to avoid damage from the start by appropriate design methods. Specially stringent earthquake resistance requirements applied to such structures are entirely justified, since maintaining them in working order can reduce to a minimum the fire damage, mass injuries, psychological traumas and infectious diseases. Special design treatment of some essential facilities, such as hospitals, are already common in some highly seismic regions. Unfortunately, a severe disaster almost inevitably affects these buildings as well, and a damage assessment may be necessary. The inspection of these essential facilities differs from that of ordinary structures in the following way:

- They get first priority for damage inspection.
- They require a Detailed Evaluation by structural engineers from the start.
- Their fixed equipment (fuel tanks, electrical panels, boilers, communication equipment, etc.) must be checked, because essential facilities need to remain operational to the maximum extent possible.
- Their fire protection system and elevators must be checked.
- Damage inspections must be coordinated with appropriate governments.

It is desirable to have a specially trained engineering team to inspect essential facilities, particularly hospitals. A preparation with review of structural drawings, seismic design calculations, soil and foundation reports, and a facility inspection made prior to the disaster is desirable.

The Kaiser Foundation, a large health maintenance organization with many hospitals in California, developed a special training program for its hospital staff (1984) [1]. Its purpose was to prepare hospital facility personnel and stationary engineers to perform an initial damage survey to determine whether to evacuate or continue operations. This plan is not intended to be a substitute for inspections by a structural engineering team, but to fill the period before the team can reach the scene. The difficulty in the use of such a program by non-structural engineers is that the damage is often in the so-called "grey area", neither safe or unsafe.

Infrastructure (roads, bridges, etc.)

As important as the facilities is the infrastructure. Roads, bridges, airports and railways are very important in making the disaster area accessible. Opening of roads and airports, repairing of bridges, building of emergency bridges, etc. are necessary to make the transportation of personnel, relief goods and material, into the area possible. Inspection and restoration of dams and rivers are necessary to avoid flooding.

Because of the emergency and necessity of these operations, they will also get priority and a definite policy to restore rather than demolish should be investigated. The repair of the infrastructure should be done as soon as possible.

Special inspection guidelines and damage classification methods for the examination of elements of infrastructure are difficult to find. We can refer to two documents.

a) "The Safety Assessment Plan for Volunteer Engineers" developed by the OES, 1988 [1] contains damage assessment report forms for eleven specific civil engineering structures, including airports, roads, bridges, dams and buildings, which utilize a damage scale from 0 to 6.

b) The "Manual for repair methods of civil engineering structures damaged by earthquakes" developed by the Japanese Ministry of Construction in March 1986 [6] proposes a damage classification for bridges (Fig. 3.21). The bridges are classified in five damage degrees based on the bearing force, it is therefore most important to determine whether or not a bridge fall will eventually occur.

No damage	There is no known anomaly concerning bearing force.
Minor damage	There is no immediate decline of bearing force.
Medium damage	The damage may affect the lowering of the bearing force, but if it does not worsen from aftershocks, live load, etc., the bridge can be used for present.
Major damage	The damage seriously affects the bearing force and may lead to critical consequences, such as bridge fall. These bridges must be totally closed to traffic.
Bridge fall	The case of a fallen bridge.

Fig. 3.21 : Five levels damage classification for bridges [6]

The report on the damage to the bridges that were surveyed by the Japanese disaster relief team during the earthquake in the Philippines of 1990 [6] includes the following data :

- Name and location of the bridge.
- Type of the superstructure.
- Type of the substructure.
- Damage conditions.
- Degree of damage.
- Remarks (available or unavailable for traffic, emergency measures taken, etc.)

Lifelines

Damage to lifelines including water, sewage, gas and electricity, is also an important element concerning safety and usability of a building. Gas leaks and downed power lines are important hazards for inspection and reuse of buildings. Water supply and sewage removal are required to avoid epidemics.

Although special inspection guidelines, damage classification systems and posting methods may be considered, little is found in the literature concerning this subject.

As mentioned in 2.7 the rehabilitation of electricity, in general supplied by overhead wire, would be the fastest whereas gas would be the slowest. Aside from the examination and restoration of the underground lifeline pipework, the major reason for the delay in the recovery of the gas supply is that the main stopcocks in all houses must be examined as a safety measure before gas can be turned on. This was proven during the recovery of the lifelines in Sendai City (Japan) after the earthquake of 1978 (Fig. 2.10 and 2.11).

3.8 Organisation

How should emergency relief be organised ? This question has two dimensions, a national and an international one. When a disaster takes place in a region, two parallel branches of relief can be distinguished. On one side there is the activity that is spontaneously developed by the population itself and its relief organisations. On the other side there is the relief that comes from outside the stricken area (national and international), normally developed after the demand and with the authorisation of the local authorities.

At the present, almost every state and even every region of the world has a kind of emergency plan and an organised structure (Red Cross, Civil Defence Force, Army, etc.) for emergency relief. However, this plan and structure varies from state to state and even from region to region. It normally depends on the development of the country (industrialised or not) and the frequency with which the region is stricken by natural disasters. So, for example, Japan will be better organised for the period after an earthquake than the European countries because it is situated on a seismic belt and is heavily stricken by earthquakes almost every year. It is also better prepared then other countries in the same region, such as the Phillipines, thanks to their bigger financial and technical resources.

Next to the local emergency organisation stands the international organisation and co-operation. This relief is normally offered after a demand from the stricken country. Usually it consists of medical and financial support but lately there is also a demand for emergency damage assessment surveys and for assistance and advice on repair, demolition, restoration and reconstruction. Japanese expert teams, for example, were asked for assistance after the earthquakes in Spitak, Armenia in February 1990 [6] and in the Philippines in July 1990 [7].

At present, there is no umbrella organisation for building technical relief and so all the requests for assistance are bilateral between the stricken country and other countries. This leads to the presence at the site of various relief teams of different nationalities. Because there is hardly any deliberation before the intervention, many organisational problems occur. Even the co-operation of relief teams from outside the region with local teams is often inefficient. After the Mexico City earthquake of 1985, for example, the demolition work done by foreign explosive experts was halted because of the local authority's lack of knowledge concerning blasting techniques (see also 4.5).

Additionally, the highly differentiated relief activities such as rescuing trapped people and dead bodies, medical care, damage assessment, demolition, tend to interfere with each other. This makes efficient co-operation and information interchange between different activities, relief teams and professions necessary.

To deal with these organisational problems and to fulfil the growing need for building technical assistance after natural disasters, the setting up of an International Intervention Team of construction engineers should be considered, in accordance with well-known medical relief organisations such as the International Red Cross or the Médecins Sans Frontière.

3.9 Recruiting and training of staff

Recruiting and training

In countries where there is the potential for natural disasters, emergency plans should be developed concerning damage assessment, repair, demolition and reconstruction of damaged buildings. These plans should also consider the recruiting and training of structural engineers, recruiting and training of volunteers for building inspectors and preventive education and preparedness of the population. This could be organised by national or local building departments and engineers associations.

Recruitment and training in the pre-disaster period is very important to make the staff familiar with the assessment procedure, the assessment forms and manuals, and to prepare the staff for post-disaster field conditions (page 95). It will also make the relief more effective and reliable. If they are not repeated, however, the benefits of the training can be short-lived.

In Japan, up till now, only the Kanagawa Prefecture as one of the local governments has the intention to recruit 6000 "Judgement Engineers for Damaged Buildings" until 1995, from the building engineers [5].

In Latin America many universities and institutions have, in collaboration with the PAHO (Pan-American Health Organisation), actively and formally incorporated disaster preparedness, prevention and mitigation into the curricula of engineering schools.

However, in the period after a disaster a lack of local staff is still likely to exist. An appropriate assessment procedure should be developed considering this. To deal with this problem, relief teams from outside the stricken area, national or international, can be asked.

In the ATC report [1] it is stated that the State of California's Office of Emergency Services (OES) has developed a plan, in conjunction with the Structural Engineers Association of California (SEAOC), to bring in volunteer SEAOC engineers from outside the damaged area. These volunteer structural engineers are used to supplement the staff of the local building department for doing the safety evaluations. OES volunteer engineers are considered to be temporary, uncompensated disaster service workers. As such they will enjoy the same immunities as officers and employees of the state and receive workman's compensation for injuries sustained on the job as provided by state law. Under state law, no disaster service worker who is performing disaster services ordered by a lawful authority during a State of War, State of Emergency, or a Local Emergency is liable for civil damages on account of personal injury to or death of any person or damage to property resulting from any act or omission in the line of duty, except one that is willful (California Civil Code).

A similar program has also been sponsored by the California State Council of the American Society of Civil Engineers (ASCE). This program provides assistance for evaluations of infrastructure, such as airports, roads, bridges, dams, water treatment and waste water treatment plants and pipe lines. ASCE also has volunteer geotechnical engineers available to assist.

Composition of the inspection team

After the disaster, building damage inspection teams have to be set up. It is important to form teams that can assess the different hazards that can occur and can handle the different problems in the field. A good example of team composition (Fig. 3.22) for the Detailed Evaluation is given by Gavarini and Angeletti [3]. This is of course only a guideline for the composition of the inspection team. It should be changed in accordance with the available staff and the damage condition of the area to be inspected. For example, in some areas where no ground failures have occurred there will be no geologist required. The proposed team composition is as follows:

1.	A structural engineer with large experience from past earthquakes, leading and organizing the team. He is obliged to inspect the building, to determine and classify the damage level and usability of the building.
2.	A geologist in charge of evaluation of possible local ground effects.
3.	An industrial engineer for the control of the mechanical and electrical systems (elevators, electrical facilities, etc.).
4.	A technician with adequate knowledge of the areas to be inspected.
5.	A labourer.

Fig. 3.22 : Composition of an inspection team [3]

For the Rapid Evaluation, less staff are required because it only consists of a quick visual, mainly exterior inspection. However, for practical and safety reasons the team should always consist of at least two people.

Field equipment

One important key to an efficient post disaster inspection is the availability of essential equipment for the inspectors. While each inspector will be expected to provide certain items that may be kept either in the car or backpack (Fig. 3.23), additional basic supplies and tools (Fig. 3.24) can realistically only be furnished by the local field offices run by the emergency inspection organizations (usually the local building departments and the state emergency services). The ATC report [1] proposes a useful list, which is a compilation of lists compiled by several organisations and by experienced damage inspectors. Of course the final composition of the field kit will be dictated by local circumstances and personal preferences.

Personal items	Field equipment
Essential items	*Essential items*
Personal identification papers	Clipboard
Official identification	Paper / notebook
Driver's license	Pens / pencils
Credit cards / cash / traveller's cheques	Flashlight and extra batteries
(including change for pay phones)	
Backpack	
Eyeglasses / safety glasses (including	
extras)	
Prescription medication	
Hard hat	
Boots / sturdy shoes	
Rain gear / extra clothing	
Personal hygiene supplies	
Dust mask	
Suggested items	*Suggested items*
Canteen	Portable battery-powered radio
Water purification tablets	Camera (possible polaroid), film,
Safety goggles	flash equipment and extra batteries
Safety vests	Cassette tape recorder, blank tapes
Gloves	and batteries
Knee pads	Tape measure
Sleeping bag	Magnetic compass
First aid kit	Swiss army knife
Sunscreen lotion	
Mosquito lotion	

Fig. 3.23 : Items provided by the Inspector [1]

Regarding these lists, it must be said that some suggested items such as binoculars, camera and lanterns seem essential; while ladders, calculators and cassette tape recorders, appear not to be very useful in achieving a quick visual assessment. The proposed list should, therefore, be discussed in accordance with the chosen assessment strategy. In general, taking too much equipment on the inspection round will obstruct the speed of the damage assessment. Depending on the circumstances the equipment should, therefore, be kept to a strict minimum.

Essential items

Street maps
Official identification / field passes
Inspection forms and manuals
Posting material (placards are colour and brushes)
Yellow "Do Not Cross Line" tape
Staple gun / thumbtacks for placards
Communication equipment (walkie-talkies, etc.)
Transportation to and from the area
Paper, pens, staples, etc.
Names and phone numbers of emergency personnel

Suggested items

Ladders
Binoculars
Battery operated lanterns and extra batteries
High capacity copier, and copy materials
Assorted tools : hammers, handsaws, crowbars, wire cutters, wrecking bars
Plumb bobs and 5 meter of line
Carpenter's levels / surveyor's hand levels
Calculators

Fig. 3.24 : Equipment supplied by the Local Authority [1]

Field conditions

After severe disasters many buildings become so heavily damaged that the danger of collapse or falling debris can be quite high. Strong aftershocks can increase this danger even more. Therefore, damage inspectors should be conscious of their own personal safety, and that of their team members at all time. The team members should be alert for possible hazards and wear appropriate safety equipment, such as helmets, safety gloves, safety shoes, masks and goggles. In case of fire or gas leaks precautions should be taken and the responsible department should be alerted. Very critical conditions, such as downed power lines and release of hazardous materials (asbestos, toxic chemicals, etc.) should be avoided.

Studies at the Disaster Research Centre at Ohio State University [1] have shown that panic does not generally follow a major disaster. However, victims have been observed to follow certain emotional patterns. Although it is not the inspector's job to provide social services, the team members should realize that the people they contact may be feeling any of a range of emotions. Making an effort to be helpful and understanding may ease the victim's mind and pave the way for quicker inspections. It is essential that the inspection team has the necessary information to direct victims to help facilities, such as emergency shelters and Red Cross field stations.

Homeowner reactions can be very different, many people contact the inspectors to get information about the safety of their homes, others are suspicious and reluctant to co-operate. These individuals should be dealt with in an objective, factual and patient manner. Posting decisions should be explained to those affected. If it is necessary, authorities should be contacted to prevent entry. To make the assessment possible it is very important that the inspection team has full legal rights to inspect even apparently undamaged structures.

Post earthquake building damage evaluation can be gruelling work for the inspector too. Long hours of work in an emotionally turbulent environment, exposure to loss, injury and possible death, along with a lack of adequate sleep or food and separation from one's family, can impact the inspector's own emotional and physical well-being. Unchecked overworking can often lead to the "burn-out syndrome", a state of exhaustion, irritability and fatigue that creeps up unrecognized and undetected upon the individual and markedly decreases his effectiveness and capability. Therefore, it is important that the field work is supported and encouraged by a good organisation. Additionally, co-workers should keep an eye on each other, and talk about their experiences and fears. They should eat well, get enough sleep and enjoy some recreation away from the disaster scene.

References

[1] Rojahn Christopher, ATC-20 report; "Procedures for Post-Earthquake Safety Evaluation of Buildings", 1989; Applied Technology Council, Redwood City, California.

[2] Rojahn Christopher, ATC-20-1 report, "Field Manual : Post-Earthquake Safety Evaluation of Buildings", 1989; Applied Technology Council, Redwood City, California.

[3]   Schuppisser S. and Studer J., "Earthquake Relief in Less Industrialized Areas.", A.A.Balkema; International symposium organized by the Swiss national committee for earthquake engineering and the Swiss society of engineers and architects, Zurich, Switzerland, March 1984.

[4]   Petrovski, Jakim, "Classification of Structural Damage and Analysis of Damage Distribution as Basis for Planning of Reconstruction after Catastrophic Earthquakes.", Skopje, Macedonia, 1979; for the second seminar on construction in seismic regions, Economic Commission for Europe Committee on Housing, Building and planning, Lisbon (Portugal) October 1981.

[5]   Kasai, Yoshio, "Classification and Judgement of Damage Rate for Reinforced Concrete and Wooden Structures", Nihon University, 23 September 1991, Rilem 121-DRG.

[6]   J.I.C.A., "Report on Expert Team of Japan Disaster Relief Team (JDR) on the Earthquake in Philippines of July 16, 1990", Japan International Cooperation Agency, August 1990.

[7]   J.I.C.A. , "Report of Japan Disaster Relief Team on Earthquake at Spitak of December 7, 1988", Armenia, USSR. Japan International Cooperation Agency, February 1990.

[8]   CEB, "Diagnosis and Assessment of Concrete Structures", State of the Art Report of the Comité Euro-International du Beton, Information bulletin N°192, Lausanne, January 1989.

[9]   CEB, "Assessment of concrete structures and Design Procedures for Upgrading (Redesign)", Comite Euro-International du Beton, Information bulletin N°162, August 1983.

[10]   De Pauw C., "Recyclage des descombres d'une ville sinistree.", C.S.T.C. Revue N°4, decembre 1982, Belgique.

4

Demolition of Damaged Structures

Erik K. Lauritzen and Martin B. Petersen
DEMEX Consulting Engineers A/S, Denmark

4.1 Description of the special condition, politics

There is a large amount of literature relating to reconstruction following a disaster - some of the most interesting titles are shown in Appendix I - however, work concerning the intermediate phase (i.e. the demolition phase) from relief to rehabilitation and reconstruction has been left out. There is very little information on emergency demolition for the purpose of rescuing people and demolition of concrete structure for reconstruction. This was very apparent following earthquakes in Mexico 1985, Armenia 1988, San Fransisco (Loma Prieta 1989), the Philippines (Luzon 1990) and Turkey (Erzincan 1992) where the rescue operations as well as the following site clearance operations caused numerous problems and loss of time.

New demolition and recycling techniques that have been developed in Japan and Europe in order to produce an efficient alternative to the traditional and often cumbersome methods of demolition should be considered more closely. The traditional methods of demolition normally have little regard to the environment and often the resulting debris is disposed of in uncontrolled sites. The extent of recycling is minimal and the waste of potential materials is immense. We would like to see a more environmental friendly attitude to this intermediate phase, be it in the form of demolition through more environmentally friendly methods or stricter control of the disposal and reuse of the arising building waste.

A major disaster has an economic effect on the local region since the loss of buildings, lifelines and infrastructure results in a slump in the local economy. It is therefore important to boost the economy by employing as much local expertise and work force as possible. This creates a unity in rehabilitation in the community and results in a more stable recovery. Due to this scenario, the demolition work should be carried out by a consortium especially set up to do the work rather than commissioning the work to individual companies. This consortium must be set up in regions of high seismic risk to ensure rapid formation after a disaster. This will combat the eventual competitiveness of the large financial investors in the community which could result in a monopoly controlled by certain individuals. It would therefore be preferable to have a local demolition joint-venture to generate the needed local income after a disaster. There will, however, be a certain need for outside managerial and consultancy aid, especially in the developing countries, and this must be acknowledged and respected. The cooperation with the outside aid must be extensive and at a high level in conjunction with the local representatives so as to maintain as much of the local culture and style as possible. The outside consultants must be cautious when introducing major resources, such as machinery, into the post-disaster phases since this may be seen as taking work away from local resources.

Demolition contractors are more frequently required to present a report on the method of demolition chosen for the decommissioning job. This is to ensure a safe and more structured demolition process and the contractor must be able to document with rational explanations the reasons for any deviations from the plan. However, in the case of a disaster there is rarely sufficient time or workforce resources to study and accept such demolition plans. We must therefore establish the required framework for demolition jobs in conjunction with the demolition consortium set up during the disaster planning phase. By setting the framework concurrently with the consortium, realistic standards can be reached and help with the rapid planning after a disaster.

The resulting plans for post-disaster phases involving demolition and reconstruction must not be so strict as to restrain the contractors from improvising. A post-disaster situation can never be fully predicted and circumstances will vary from site to site. This reinforces the need for there to be local decisions made in each community, these decisions lying inside the framework of that set out under the disaster planning

Fig. 4.1 Photo of collapsed double-deck highway bridge, Interstate 880 in Oakland, after the 17th October 1989 Loma Prieta Earthquake.

phase. The unpredictability of the post-disaster situation needs quick reliable decisions made on-site, which do not need to go through bureaucratic processes. Each disaster must therefore be covered broadly and with specific frames.

It is thought relevant to study the feasibility of preparing studies for all infrastructures located in areas considered to be high risk in terms of earthquakes and other disasters in order to compile a post-disaster reconstruction programme. These programmes would relate to specific local communities as a part of the regional area affected by the disaster. Such a study would include details of the building materials utilized in the area, the building types, the available work force in the field of demolition and other related factors. These programmes are kept updated by the local community and are kept accessible for use after a disaster.

In this manner we are able to prepare a region for the recovery phase of a disaster which is about to happen. The programme would also hold information on link recovery areas, such as details on open areas which could be used for recycling plants and lifeline analyses for the opening of such services as electricity, roads and gas.

The type of organisation implemented after a war is usually different since in this case the area affected is normally of a national or a conti-

Fig. 4.2 Photo of collapsed building, Hyatt Hotel in Bagdio after 15 July 1990 Earthquake.

nental scale and not local as with most other disasters. Therefore such factors as recycling plants and dump-sites must be planned differently to cope with the distances and current state of stability of the country (ie. where to start the clearing, who should carry out the clearing of damaged structures etc.). In these cases, local communities would be given managerial and consultancy aid in order to decentralise the decision making process. This speeds up the recovery phase and results in greater community satisfaction.

Before any demolition of any type is employed in an area, it is vital that the rescue phase has ended completely. The rescue teams must have given clear information to the contractors that their rescue phase is finished in the selected area since any demolition work carried out may reveal survivors. Such situations are highly sensitive and must be respected.

A pilot project for the recycling of building waste after the Algerian earthquake, 1980, was halted due to several reasons, including political setbacks. One of these reasons involved the local population's refusal to live in buildings with risk of eventual remains of bodies in the walls [1]. Other political complications are present when the demolition contractors wish to use explosives for the rapid demolition of damaged buildings. In Mexico City after the September 1985 Mexico Earthquake the use of explosives would have minimized the hazards to labourers working on demolishing buildings by introducing more rapid and safer working environments. The political decision was, however, taken that explosives could not be used due to the fear that eventual survivors in the rubble could be harmed and bodies could be further buried [2].

The larger the area of demolition and clearing sites, then the more economical the process of demolition and recycling becomes. For massive scale demolition and clearing processes it is possible to use less sophisticated methods of demolition which are often quicker and thereby more economical. These methods, such as the use of wrecking ball and explosives, may not be appropriate in certain instances where sensitive surrounding buildings and installations exist. However, if a whole area was made a demolition and recycling site, it would be possible to utilize these methods without detailed concern for the immediate surroundings which are also part of the demolition process. The use of Expropriation Decrees to evacuate a site is a sensitive issue with the local population for obvious personal reasons, but must be considered. This may apply to an area of severely damaged buildings where it may be more economical and quicker to select total demolition, even if there exists a few not so severely damaged buildings. These less severely damaged buildings would then be demolished rather than being rehabilitated. This case can also be attractive when selected buildings are chosen for rapid demolition in order to stop the spread of fires arising from disasters. Here the buildings are blasted to avoid the spread of a fire from one local area to another local area.

With the formation of a large area of demolition, it is possible to organize mobile crushing and sorting plants for recycling of the building waste, thereby cutting the transportation costs and logistics problems which arise from transporting material in the region.

The rapidity with which the project can be established is important to the procedures of the project. A long delay will increase the possibility of the disaster struck population inhabiting the ruined buildings and using them as temporary shelters, thereby increasing the dangers should such buildings collapse due to after shocks. This temporary inhabitation can also inhibit the demolition process, as was the case in El-Asnam, Algeria.

The rapidity with which the demolition project can start is restricted by numerous factors, many of which will not become apparent until after the disaster. In this time period many previously unaccounted for factors will arise delaying the implementation of the demolition phase. However, with the local authorities having established an advance view for rescue action and procedures, the implementation of an effective primary demolition phase is possible. So we must deal with the apparent problems as early as possible in the disaster planning phase.

Several of the potential delays in activating the rebuilding phase arise from above mentioned factors, such as decision making chains, establishment of demolition consortiums and determining realistic working frameworks.

Referring to the decision making process after a disaster, it is clear that due to the circumstances clear and well defined resolutions are difficult to establish. Therefore minimal need for important judgements is attractive, this being attained through careful planning. With a decentralization of the political structure, each local authority would be able to control their own situation, thereby giving the local population a say in the phases. Without having to refer to a distant city a team of specialists, chosen before the disaster as stand by, would be able to aid the local authorities in utilizing the pre-planned post-disaster phases. The smaller and more specialized this team the better, since numerous team members may confuse the issues.

As mentioned before, the establishment of a demolition and recycling consortium would simplify the tendering of work after a disaster. Set guidelines for the imminent demolition work by the consortium would be controlled by a team of engineers, who may also work in conjunction with the rescue phase in order to help with lifting and demolishing of elements trapping humans. It is advantageous to involve the demolition engineers as early as possible, in this way the team of engineers will also gain an appreciation of the scale of demolition needed and the possible techniques there could be used. There would be close cooperation between the demolition engineers and the structural engineers who will be assessing the affected buildings for severity of damage. It must be appreciated that the reconstruction phase for the consortium allows for the loss of resources by the contractors due to the disaster.

The evacuation of buildings is paramount, especially after earthquakes where after shocks can cause fatalities, and must be definite before further action can be taken. The method of clearing a building for demolition is to be strict in order to ensure no misunderstandings, a procedure which is to be defined in the disaster planning. This reinforces the fact that communication between the different relief teams is important, why decentralization is important so as not to complicate and delay these communications.

The approved plans agreed upon by the local authority in the disaster planning phase may seem reasonable before the disaster, however, afterwards local opinions may be negative. This will be apparent in the cases of assessment of buildings, expropriation and demolition. Therefore it is important to try to reach an agreement with the local population from the start. Representatives from the local community should be allowed to join the decision making board.

The conclusions are that it is important to plan ahead for a post-disaster phase handling programme which should set guidelines for the difference phases. Also the guidelines should not be restrictive and open to improvisations to be approved by the local specialist team, thereby keeping the process running and in local control.

4.2 Demolition methods and machines

Demolition in connection with clearing and rehabilitation projects may be divided into the following main categories:

- Complete demolition
- Partial demolition

Complete demolition involves the removal of the structure in question while partial demolition represents a limited measure. As a basis for the evaluation and choice of working methods for the demolition of buildings and structures, a certain knowledge is required as regards the relevant materials and the technical possibilities of demolishing them. A special regard should be paid to the instability of the structure due to the extent of the damage.

A variety of methods can be applied to the demolition of concrete buildings; they can be divided into the following principal groups:

- Crushing
- Chopping
- Splitting
- Blasting with explosives or chemical agents
- Cutting and drilling

For the benefit of the disaster scenario, then we have to determine the method of demolition to be implemented. This will ordinarily be influenced by three factors:

- Time
  Is the building close to collapse and therefore dangerous to the surrounding environment? Is the building a link for the spread of fire?

- Structural stability
  Is the collapsed structure unstable? And is it possible to support the structure and render it safe during demolition?

- Sensitivity
  Does the building contain hazardous material?
  Are the surroundings sensitive?

The above mentioned factors will thereby determine the method of demolition, be it an unsophisticated method for a rapid demolition or a finer more precise technique for the sensitive cases.

In connection with emergency housing shelters, it might be possible to use large portions of the elements from damaged buildings to construct temporary housing with simple and rapid techniques. In this way the undamaged concrete elements can be used, however, this depends on the accessibility to lifting equipment immediately after the earthquake.

Fig. 4.3 Photo of demolition of earthquake damaged double-deck highway in San Fransisco 1991.

4.3 Traditional demolition methods

The most common forms of large scale coarse demolition involve the mounted hammer, wrecking ball and the use of explosives. Dependent on the availability of the heavy machinery needed and the accessibility to the building in question, it is often relatively easy to decide upon the method to be used. For all types of coarse demolition, it is important that the building to be demolished is separated from the surrounding structures, thus easing the process and making execution of the demolition quicker. If the building in question is severely damaged and there is difficult access for machinery due to rubble or geographical barriers and no sensitive surrounding buildings exist, then blasting may be considered favourable. This obviously requires a qualified and experienced explosive engineer on site, having been included in the disaster planning phase.

Crushing with a ball and chopping with a hydraulic hammer or a manual hammer are methods used in most countries. However, these methods may have a serious impact on the structure and might cause danger to the working personnel and others due to risk of falling materials from unstable structures. Both techniques require experienced operators, especially in post-disaster situations where many unstable surrounding buildings exist. However, these two methods are rapid and efficient techniques for the total demolition of buildings.

The use of a *wrecking ball* involves the repeated impact of the steel ball, normally 0.5 - 2.0 ton, which is suspended from a mobile crane. The ball is either swung sideways with impact against the side of the building or member or is dropped onto the member from a height. The use of the wrecking ball is a rapid method requiring no prework and is suitable for all types of building materials on a large scale. The resulting building waste material is however often of medium portions and therefore secondary demolition may be necessary to reduce its size.

With a mounted *hydraulic hammer* a stable working platform is needed for the vehicle, which may be difficult in post-disaster situations. The repeated impact of the hammer at one point with shock waves progressively cracking the concrete results in relatively small size building waste rubble which is more easily accepted by the recycling plants. The noise and dust levels arising are however worse than that of other demolition techniques. The vibration levels can also be damaging to surrounding concrete. Suitable protection of the surrounding environment must be established to ensure minimal damage.

A mounted hydraulic breaker, which crushes the concrete between a solid C-shaped frame or strong jaws, also results in smaller sized portions of building waste after demolition. The C-shaped frame may be suspended from a crane, whilst the strong jaws can be mounted on a crawler type vehicle. From an environmental viewpoint, the breaker is more friendly than the two previous mechanical methods and the low levels of noise and vibration, makes this technique for demolition popular in urban areas. The availability of this machinery in some developing countries may be restricted due to the high purchase and operating costs.

Blasting requires prework in the form of drill holes bored to precise specifications. This drilling phase will influence the maximum level of vibration and noise experienced with this technique, since the actual blasting is controlled to minimise vibrations and is instantaneous to cut down on noise.

The blasting of a building quickly eliminates the possibility of the building collapsing due to after shocks, internal strength deficiencies or inhabitants. With short preparation times, few days for planning, preparing and blasting, the total demolition is relatively rapid and does not risk labourers on or near the building with handheld equipment. The resulting size of the building waste material is medium and may therefore need further demolition (ie. by a hydraulic hammer or wrecking ball) in order to reduce the material size to more manageable dimensions.

Supply of explosives, detonators and equipment may come direct from military supplies and cooperation between contractors and the military engineers is desirable in order to share expertise.

However, there will always be problems with the use of explosives in some countries since private use by contractors is not permitted. The use of explosives may solely be operated by the military, thus causing barriers for the explosive engineers. This must be negotiated in the planning for the post-disaster programme and cleared before action can be taken. Other possible barriers to the use of explosives involve the local populations fear of trapping survivors or dead bodies of loved ones by the collapsing of the building. This is obviously understandable and refers back to the communication link with the rescue teams having cleared a building for demolition.

The choice of demolition for total construction clearance after a disaster is heavily dependent on the available resources in the locality. The demolition and reconstruction phases would not be implemented until several weeks after the disaster due to rescue operations and lifeline restoration having first priority. Therefore the accessibility to machinery will increase with an increase in time after the disaster, also enabling outside help to deliver resources. Some of the techniques which will be chosen in the demolition phase may be available from the local workforce, thereby requiring outside expertise, a subject already dealt with.

The availability of *heavy lifting machinery* has highest priority for the rescue phase in connection with removing fallen members trapping survivors. This machinery will then become accessible to the demolition engineers.

Therefore, the more machinery required for a demolition technique, the longer the delay will normally be for the execution of the demolition. Thus a technique, such as blasting, which requires minimal machinery is favourable. This high degree of mobilisation is important for the post-disaster phase.

4.4 Refined demolition methods

Partial demolition is carried out in connection with the rehabilitation of concrete structures which may be damaged. The lightly damaged building is categorised and the members to be replaced or treated are identified. This may, for example, involve the removal of supporting beams for replacement or the exposure of reinforcement bars for repair work. The number of techniques for localized removal or cutting of concrete is numerous and dependent on the accessibility, availabilty and workforce skills. For example, *blasting* is very favourable, e.g. mini-blasting, but requires a special license, whilst handheld hammering is common but environmentally damaging. The mini-blasting method is developed by DEMEX Consulting Engineers according to Lauritzen [3]. Experiments involving careful blasting using small explosive charges have also been performed in Japan [4].

Partial demolition is usually expensive and therefore calls for special attention when choosing the appropriate method of demolition and the planning for the implementation so as to work in consideration for the project and the environment.

Cutting and diamond drilling are feasible considering the impact on sensitive and unstable structures, however the method takes time.

The abrasive water system projects a water jet from a nozzle at twice the speed of sound with a water pressure generated of 280 MPa to 380 MPa depending upon the waterjet device used. With a discharge volume of up to 21 litres/min, this water jet containing an abrasive material, most common being garnet or steel particles, can cut through reinforced concrete. In a paper presented at the second RILEM Symposium on Demolition Methods and Practice in 1988 [5], K. Konno of Taisei Corporation describes experiments on cutting speeds and depths in reinforced concrete specimens of 21 MPa with two layers of reinforcement bars 16mm in diameter. With a cutting speed of 2 cm/min, a depth of 50 cm can be accomplished, whilst at 10 cm/min the first layer of reinforcement bars cannot be cut and the depth remained at

Method	Primary Application	Disadvantages
Breaker, handheld	Crushing of thin walls, brackets and floor slabs in connection with reparation and rebuilding, used where access and working conditions are poor and strict environmental standards set.	Limited cutting thickness and range, unsuitable where the reinforcement bars are to be retained. Heavy equipment, best if supported on tackle or the like. Use of face mask necessary.
Breaker, mounted	Demolition of concrete columns, beams, balcony walls and floor slabs in connection with environmentally sensitive projects. Partial demolition of concrete.	Cutting of reinforcement bars can cause difficulty, not suitable for work where bars are to be retained. Use of face mask necessary.
Hammering, handheld	Cleaning of demolition boundaries in connection with partial demolition and reparation. Exposure and cleaning of reinforcement. Other minor concrete demolition tasks.	This method is expensive as it causes much noise, dust, vibrations and physical damage to user. Must use face mask, ear plugs and respiratory equipment. Danish recommendation pr. day: hour.
Hammering, mounted	The larger machines apt for larger projects inside a suitable range. Smaller machines more appropriate for minor tasks in reparation and rebuilding of concrete structures.	Hammering involves environmentally damaging aspects including dust and noise; larger machines also vibrations. Access must be large enough for machine. Remote controlled equipment recommended to reduce hazards, ear plugs and face mask necessary.
Bursting, explosives	Demolition of massive non-reinforced concrete structures and in environmentally cautious areas.	Requires pre-work with diamond boring machine. Crack development is difficult to control.
Blasting, explosives	Holes in concrete slabs more than 30 cm thick. Demolition of reinforced concrete in large quantities. Mini-Blasting for reparation and rebuilding, and the exposure of reinforcement bars, where the bars must be used again for recasting, eg. concrete columns and brackets.	This work requires special education and licenses. Some work to clean fracture boundaries with hand-held hammering or water jet is necessary after blasting.
Blasting, non-explosives	Demolition of larger concrete structures, eg. non-reinforced foundations.	Considerable reaction time is needed for agents to expand properly. The chemical reaction necessitates personal protection.
Cutting and drilling diamond	Holes in concrete slabs. Demolition work where clean boundaries are necessary. In combination with other methods.	High noise levels and water refuse.
Cutting and drilling fuel oil flame	Cutting and drilling of strong reinforced concrete.	Requires special education and experience. Fire risk.
Water jet	Surface treatment of reinforced concrete. Used for the removal of layers, drilling and cutting.	Requires certain safety regulations. Considerable water refuse. Equipment should be mounted. High risk and physical loads if used hand held.

Table 4.4 List of demolition methods and their applications.

the 20 cm. Set in practice, this resulted in walls of 15 cm in thickness, being cut at a cutting speed of 8 cm/min, which proves to be an efficient speed for the removal of concrete blocks.

There is however the problem of leaving the reinforcement bars undamaged as in many cases of partial demolition, and here the abrasive water jet is not appropriate in controlling the state of the bars if a rapid cutting speed is selected. The use of only a water jet without the added abrasives will just cut the concrete without damaging the reinforcement bars, however, this is not as efficient as with an abrasive.

Expanding chemical agents might be useful, but the drilling might cause vibrations, and the expanding time is often rather slow, often 10-20 hours.

The demolition methods which fall into this group of techniques may need further demolition since the member to be replaced will only be cut from its surroundings and removed, making way for the replacement. The removed portion will then be located in an area where it can be demolished safely. This dismantling process is very favourable if lifting equipment is available. In this way it is possible to remove layer by layer of the either collapsed or severely damaged building. There are various methods of localized cutting the concrete and reinforcement bars, including mini-blasting, diamond saw, water jet and handheld hammers. Consideration must be given to the condition of the building in order to decide the safest and most efficient method of cutting. Large and cumbersome machinery may be difficult to bring to the operation site and use of the machinery restricted. The instability of the building must be determined, since tilting will cause problems when cutting the reinforcement bars. The release of the affected portion must be carefully planned and executed since the movement of a freely suspended concrete beam can be difficult to control.

An example of the problems which can arise due to the tilting of a building is given in "The Royal Engineers Journal" paper on Earthquake Relief in Mexico City [6]. Here the top four floors of a six storey building had collapsed onto the remaining lower floors and the resulting rubble had to be removed layer by layer to prevent further collapse. With the extreme tilting of the building it was necessary to abseil down the side of the building and cut the reinforcement bars with oxy-acetylene torches. This supports the need for a mobile localized cutting

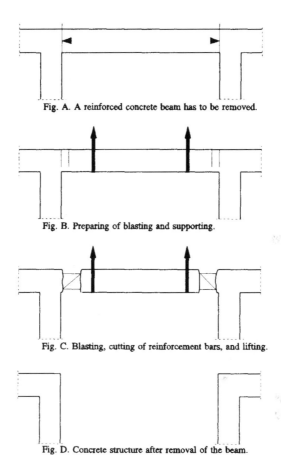

Fig. A. A reinforced concrete beam has to be removed.

Fig. B. Preparing of blasting and supporting.

Fig. C. Blasting, cutting of reinforcement bars, and lifting.

Fig. D. Concrete structure after removal of the beam.

Fig. 4.5 Principle of dismantling using the "mini-blasting" technique.

technique which is not dependent on large machinery. A combination of Mini-Blasting and oxy-acetylene torches is favourable, whereas a diamond disc saw would be difficult to handle.

4.5 Careful and rapid demolition in rescue operations

The principal choice of methods will depend on the situation of the relief operation. In the case of rescue operations or emergency relief, the time taken and the impact on the structure will be the decisive factors, because the risk of injuring trapped people or causing collapse of any unstable structures must be kept to an absolute minimum.

Blasting to create holes in reinforced concrete slabs

In connection with the demolition of a 6 year old heavily reinforced concrete bridge tests were carried out, using various techniques (including blasting), to establish the best way of creating holes in an 800mm thick slab.

The objective of these tests was to determine the time for penetration through the bridge deck in the case of an emergency rescue operation by the different demolition methods.

In the blasting tests, two methods were used. Careful blasting, which involves two or more independent shots, and rapid blasting involving one blasting with millisecond delays between the shots. Figures 4.6 and 4.7 illustrate the drill-hole patterns employed for the blasting of the holes in the bridge deck.

The tests for rapid hole blasting showed that with one blast it was possible to penetrate the bridge deck with a suitable hole (Figs. 4.8 and 4.9). For the blasting the two innermost rows of hole were charged with 125 g per hole, and the outer rows with 185 g per hole, making a total charge of 3720 g. Delay between the innermost holes and the outer rows was 25 ms. Square holes of approximately 1.5 x 1.5 m were blasted and the total volume of concrete removed was 1.8 m^3. The specific charge was 2.1 kg/m^3.

The results from blasting vibration measurements are shown in Figures 4.10, 4.11 and 4.12. With regards to the blasting effects to the surrounding concrete, simple arithmetical rules apply, ie. cube root calculations relating to a spherical blasting medium, and square root calculations to material in a single plane medium. We are thus able to combine the two variables, distance (d) and charge size (L) in a scaled distance ($d_s = d/L^{-2}$) and thereby present the results in a two dimensional graphical display. This gives an impression of the blasting effects to the surrounding material.

It can be seen that the greatest effects originate from blasting in the bridge deck which corresponds to the fact that these blastings involved a considerably larger amount of simultaneously detonating explosive charge (max. 15 kg distributed in two ms-delay steps). The results from the blasting of the holes in the deck show that a considerable reduction in the damage caused to the remaining construction is obtained with careful blasting, that is to say by dividing the blasting operation in two separate blasting sequences, [7].

Chopping

Chopping with a Montabert 1100 (weight 1350 kg) hydraulic hammer also caused perceptible dynamic effects on the bridge. The most severe effects were observed during chopping of the deck, whereas chopping of the edge beam caused less vibrations. The results of the recordings during chopping showed maximum acceleration, approx. 110 m/s^2, velocity 100 mm/s, and displacement 0.1 mm.

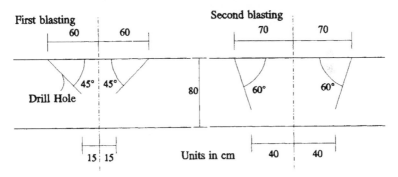

Fig. 4.6 Diagram showing drill-hole pattern and charge positions for careful blasting of hole in area C. The resulting hole was clearly rectangular 1.3 m x 1.8 m. Total charge size for both shots was 2.64 kg, and 1.872 m^3 concrete was removed, thereby giving a specific charge of 1.4 kg/m^3.

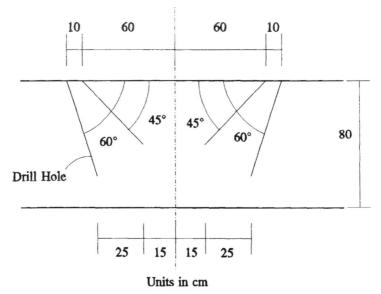

Units in cm

Fig 4.7 Diagram showing drill-hole pattern and charge positions for rapid blasting of hole in bridge deck.

Fig. 4.8 Photo of rapid blasted hole in a 800 mm thick concrete slab.

Fig. 4.9 Photo of rapid blasted hole from beneath.

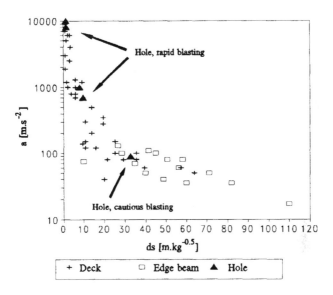

Fig. 4.10 Maximum registered acceleration with blasting of deck, edge beam and holes.

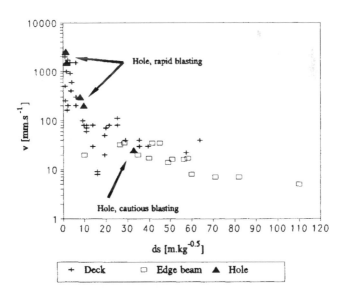

Fig. 4.11 Maximum registered velocity with blasting of deck, edge beam and holes.

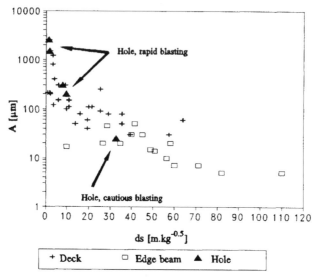

Fig. 4.12 Maximum registered displacement with blasting of deck, edge beam and holes.

Energy spectrum density and Power spectrum density

The energy or power which passes through a given point in the concrete deck can be found by summation of the Energy spectrum density (ESD)/Power spectrum density (PSD) of the whole frequency pattern. Below is given a brief account of the ESD results and their interpretation.

ESD-spectrums are calculated from the measurement results of blasting tests in the bridge deck and edge beam, as well as chopping by hydraulic hammer. It is also necessary to compare the effectiveness of each demolition method when evaluating the energy transmissions. The blastings result in the crushing of varying sized fragments, whilst the energy levels from the hammer give single chops. Therefore it was necessary to carry out multiple impact calculations of the relative energy associated with the volume of demolished concrete.

Due to the many uncertainties associated with the formulae used, it is advised that one does not place too much weight on the results obtained. Under these circumstances then it is proposed that much more research is carried out in this field in order to establish a much broader understanding on the effects of the demolition methods in relation to "cleaner technologies" strategies.

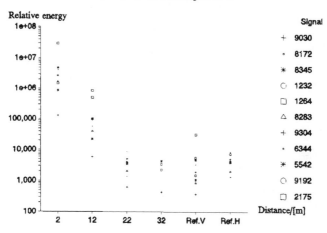

Fig. 4.13 Relative energy registered by measuring points with bridge deck blasting.

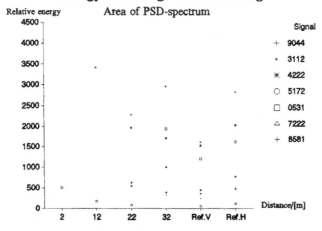

Fig. 4.14 Relative energy registered by measuring points with edge beam blasting.

Energy from hammer and ball

Area of PSD-spectrum

Relative energy

	Legend
2000	
1800	+ Ball/Ham/Bh2.1
1600	· Ball/Ham/Bh2.2
1400	✳ Ball/Ham/Bh3
1200	○ Ball/Ham/Bh4
1000	□ Ball/fresh conc.
800	△ Ball/poor conc.
600	+ Ball/fresh conc.
400	· Deck/Ham/Hb
200	✳ Deck/Ham/Ha
0	○ Deck/Ham/Hc

Distance/[m]

2   12   22   32   Ref.V   Ref.H

Fig. 4.15 Relative energy registered by measuring points from single blows with hammer and wrecking ball.

Figures 4.13, 4.14, and 4.15 show relative energy levels calculated according to recordings of vibrations (signals) at different distances from the demolition sites on the concrete bridge deck. The measurement points were established in a chain like seismic recordings with distances of 2, 12, 22, and 32 m from the demolition site (site of impact). At a fixed point at the deck vibrations were measured in vertical direction (ref. V) and in horizontal direction (ref. H).

From the figures it can be seen that there is a considerably greater damping effect with blasting than with chopping. It was also noticed at the reference point, that the horizontal effect from blasting is generally larger than the vertical.

With blasting in the bridge deck, it is seen that both the larger and smaller charges result in energy of similar magnitude at the furthest of the measuring points and at the reference point. This could be a result of greater damping of the larger charges than of the smaller charges.

The energy flow, registered at the four measuring points of the chain, is used as a basis for the comparison of the energy which is distributed in the construction as a result of blasting and chopping. This energy flow must be proportionate to a specified length of edge beam or volume of bridge deck demolished by blasting or chopping.

As a result of an evaluation of the relative energy which is involved from the dynamic effects acting on the structure, it is possible to construct a comparison of the different demolition techniques. This will help in making the decision as to which is the most appropriate demolition technique.

Protective covering for the blasting comprised of rubber mats, Fibertex and Robtex (Kevlar - Aramid fibre) mats with a 1 - 1.5 m layer of gravel. No covering was used under the bridge due to the vertical fall of the flyrock being taken by the underlying ground. Additionally, any vertical flyrock would not be able to escape the sides of the bridge.

Both blastings for holes resulted in well defined and cleanly cut holes. The difference between the time taken for rapid blasting and that for careful blasting was two hours extra for the careful blasting. This is mainly due to the fact that the covering and cleaning operation had to be carried out twice, however, the vibrations from the careful blasting are considerably less.

Rescue work

The heavy machinery that can be used for demolition and clearing is also useful for the act of rescue work. This machinery is needed immediately after the earthquake for lifting and aiding pedestrian rescue workers, and as soon as the rescue phase has come to a close, they can step in and start clearing the area. It is accepted that the most important phase is to get the survivors and dead bodies out as soon as possible. But there should also be a small group of experienced engineers who can start to classify the buildings as to whether they should be demolished, rehabilitated or sustain no damage. This improves the speed with which the demolition and clearing process can start and thereby the recovery of the region.

Blasting is an effective, safe and efficient method of preventing a building from collapsing uncontrolled as an effect of after shocks, or in the connection with demolition. With the use of blasting, the working environment is improved, especially in relation to demolition work being carried out by chopping or other manual operations. The local authority's lack of knowledge regarding blasting techniques has restricted its use for the clearing of buildings after earthquakes. This can be seen after the Mexico City earthquake of 1985 where the authorities halted blasting work [2], due to the possibility of there being survivors in the damaged buildings. Blasting was not used in El-Asnam, because the survivors quickly began to use the damaged buildings as temporary shelters.

The quality and quantity of the materials, and thereby the possibilities of reuse in new constructions, is conditional upon the quality of the existing concrete and the amount of "pollution" in the crushed concrete.

4.6 Organizing, education of people to assist the demolition work

In many countries, this includes several EC countries, USA and Japan, the crushing of building waste and the reuse of materials for filling is routinely carried out by demolition entrepreneurs. These recycling activities are carried out in accordance with the entrepreneurs own initiative, due to a cost related evaluation of the given conditions concerning disposal of the building waste without real regulation or coordination from the authorities. In The Netherlands and Denmark it was found necessary to perform control and coordination of the waste handling inside certain geographical regions.

Since the recycling of building waste is especially directed at supplementing the raw materials, then it is appropriate for the authorities to control and coordinate the recycling activities from a complete evaluation of raw material management in the geographical region concerned.

It should also be mentioned that demolition work is risky and should be performed by skilled people under technical supervision. After natural disasters or war operations, there is an immediate demand for rescue operations involving demolition work and clearing of the building materials in the search for any eventual survivors or bodies which may be in the ruins.

After this rescue phase, the subject of rebuilding should be discussed at the same time as the political and financial aspects of rebuilding are set. Appropriate evaluations of the different types of damage and

Fig. 4.16 Inadequate demolition on an earthquake damaged building site in Bagio 1992, $1^{1/2}$ year after the Luzon earthquake in 1990. The people are trying to get a wall down in order to take out the RF-steel bar for scrap metal.

materials involved should be carried out as soon as possible in order to ascertain quantities and dimension the recycling system so as to require the optimal recycling equipment for that particular situation. This will incorporate the necessary size and types of crushing plants needed.

The strategy for the system is set according to:

- Time aspect of the duration of the rebuilding phase.
- Need for raw materials in the new constructions and possible fill in the region.

Before the recycling system is established, and the handling plants are set up, it is appropriate to organise the rescue operations as much as possible to carrying out selective demolition and sorting of the building waste materials. It is also necessary to undertake rapid evaluation of the demand for emergency shelter and the possibilities of using materials from the damaged buildings for these shelters.

As the recycling system is organised, concurrent with the rebuilding of the area, it is important that it is carried through according to the above named principles for permanent regional plants. The system should provide the most effective clearing and best possible supply of raw materials for the soil improvement and new construction.

Specific details from recycling in connection with larger clearing projects are not yet available. However, after the Algerian earthquake in El-Asnam in 1980, CSTC (Centre scientifique de la Construction, Bruxelles), with the support of the Belgian government, undertook a pilot operation involving recycling of building waste [1]. The project was later halted due to political reasons.

After the Armenian earthquake in 1988, crushing plants were installed in 1990 in connection with the clearing work. Results from this recycling plant are not available yet and are still being collected.

It is in the meantime considered important to develop further evaluations concerning the possibilities for a rational handling and recycling of building waste as a result of catastrophes.

The recycling of building materials is evaluated as being an important aspect to the International Decade of Disaster Reduction, where one of the work areas involves the relief of impact on major cities [8].

The Japanese are very advanced in this field and have set up several "courses". For example, a local government in Kanagawa Prefecture intends to admit 6,000 on a course titled "Judgement engineers for damaged buildings" [9]. Another course in "Construction Control Engineer for Demolition of Building" has already been implemented (1991).

The quality of the local work force must be considered in connection to the production of materials for reuse, when the different techniques of demolition and equipment are chosen. This quality will affect the amounts of recyclable material produced as seen in the pilot project in El-Asnam, where the different types of demolition resulted in different amounts of material.

Type of building	Recyclable materials (x %) depending on type of demolition		
	A-Demolition by experts and modern techniques	B-Demolition by experts and moderate equipments	C-Demolition without any special equipment
I-Dwelling, Multi story	80 %	60 %	40 %
II-Houses single story	30 %	10 %	0 %
III-Commercial and industrial buildings	70 %	50 %	30 %

Table 4.17 Recycable materials depending on type of demolition and building [1].

The above table illustrates the influence of the local El-Asnam construction traditions on the amount of recycled material produced. The single floor buildings were commonly built by poor quality masonry or mud bricks where a minimal amount of the building is recyclable. The multi-story buildings were typically poor quality concrete constructions, leading to larger amounts of materials for recycling. In regions of different building traditions and materials, the above amounts would naturally be of correspondingly different values.

4.7 Case Study, the Erzincan earthquake,Turkey, 1992

After the earthquake of the 13th March 1992 a lot of damaged buildings had been demolished and the debris transported to an emergency dump site approximately 10 km outside the city.

The disposal site was visited the 12th August 1992, [10], and the amount of mixed rubble was roughly estimated to be in the region of 500,000 - 1,000,000 t which was tipped over a large area. On the road to the dump site it was noticed that a lot of building waste and other types of solid waste were tipped uncontrolled along the road.

Several damaged buildings were inspected, amongst which

- A number of buildings under construction before the actual earthquake were noticed in the western part of the city.
- A 6 story building with a supermarket in the basement.
- The hospital, where a Danish rescue team had worked.
- One remaining 4 story dwelling block from a total of 20 others which had been demolished and removed.

A lot of damage had occurred, but the city had succeeded in re-establishing normal life and removal of a large part of the total damaged structures, in the few months since the earthquake. However, a lot of the destroyed structures still remained to be demolished and removed, and huge amounts of different kinds of solid waste had to be disposed of.

The following issues were recommended for implementation in the Rehabilitation and Reconstruction Project as post-disaster phases:

1) An Environmental Assessment (EA) covering necessary aspects of the environmental protection connected to the Rehabilitation and Reconstruction Project.

2) Based on the EA and other requirements, the possible sites of the sanitary landfill and sewage treatment plant are decided upon.

3) A masterplan for all solid waste including construction and demolition waste, (C&D waste) should be prepared in accordance with Turkish legislation (e.g. regulation of solid waste March 1991) to prevent further inadvertent tipping of building waste (see Figs. 4.18 and 4.19) and other uncontrolled waste disposal.
   (It should be noticed that in several European countries, as well as in the USA, mixed C&D waste is not considered as inert waste due to leachate. Mixed C&D waste should be disposed of in sanitary landfills.)

4) Demolition of remaining damaged structures and buildings and clearance of all sites with remains from buildings should be planned and organized in order to reduce the total costs and the time taken. Selective demolition and use of handheld or mounted hydraulic concrete breakers is recommended with respect to separating concrete and reinforcement steel on the site. Modern demolition

techniques with adequate machinery and probably blasting techniques are recommended with respect to safety and efficiency of the work performance. All organic material, PVC, paper etc. must be sorted from the inert waste material.

5) Recycling of steel and concrete should be encouraged so the amount of waste to landfills is minimized. The recycling work should be incorporated within the demolition work. Steel should be collected and prepared for further treatment on scrap yards.

6) Based on the results of the analyses of the drilled concrete core samples the potential reuse of crushed concrete rubble should be investigated carefully. It is realized that the area is rich in natural resources, however, the crushed concrete will be suitable for many purposes e.g. base course materials for roads, backfill etc.

7) The damaged reinforced concrete frames of buildings under construction before the earthquake in the western part of the city should be demolished by blasting or by the use of concrete breakers (see fig. 4.19).

8) The collapsed building structure (see Fig. 4.20) is specifically suitable for relief experiments and training of rescue teams with respect to investigation of the most applicable methods of rescuing trapped people.

9) Due to the lack of tents, emergency procedures and plans should be prepared concerning the establishment of shelters and improvised houses using materials from damaged buildings etc. Moreover, the possibilities of planning the provision of emergency houses made by prefab elements, for instance the houses presently established for the building workers (see Figs. 4.21. and 4.22), should be discussed.

10) The urban vulnerability as mentioned in the World Bank report Annex IV page 37 should be combined with the EA (see 1), and it is recommended that a standardized approach and methodology based on common risk analysis techniques is used.

Fig. 4.18 Reinforced concrete tipped on an area east of Erzincan

Fig. 4.19 The road from the city to the dump site. In the background tents for emergency housing.

Fig. 4.20 Concrete buildings during construction before the earthquake of which several were damaged.

Fig. 4.21 Collapsed 4 storey dwelling building.

Figs. 4.22 and 4.23 Improvised emergency housing substituting the damaged buildings in the background.

References

[1]   C. De Pauw, "Recyclage des Decombres d'une Ville Sinitree", C.S.T.C. - Revue/N°4/ Dec 1982, Bruxelles, 1982.

[2]   Richard A. Dick, "S.E.E. Members Aid Mexico in its Time of Need", The Journal of Explosives Engineering Volume 3, Number 4, Nov./Dec. 1985, The Society of Explosives Engineers, Dublin, Ohio.

[3]   Erik K. Lauritzen and Martin B. Petersen. "Rehabilitation & Repair of Concrete - Partial Demolition by Mini-Blasting". Concrete International, June 1991.

[4]   Yoshio Kasai. "Report on the experiments for protection methods from flying objects in the demolition of urban structures by blasting". All Japan Association for Security of Explosives, 1990.

[5]   Y. Kasai. "Demolition Methods and Practice", Proceedings of the Second International RILEM Symposium, Tokyo, Japan. Chapman and Hall, London, UK 1988.

[6]   Major D M Webb and Captain P D Cook, "Earthquake Relief in Mexico City", The Royal Engineers Journal Volume 100, No. 4, December 1986, Chatham, Kent, UK.

[7]   Erik K. Lauritzen, Niels K. Madsen and Jens Jensen. "Demolition of Motorway Bridge, Great Belt Link - A Research & Development Project on Fragmentation and Recycling of Reinforced Concrete 1991-92". DEMEX Consulting Engineers A/S. Denmark, 1992.

[8]   Dr. Ewald Andrews. "Bevölkerungsschutspolitik im Nord-Süd-Dialog: IDNDR 1990 - 2000". Bundesverband für den selbsschutz, Postfach 20 01 61, 5300 Bonn 2, 1992.

[9]   Yoshio Kasai. "Japanese Working Group for the RILEM 121-DRG, Task Force 2 Report", Nihon University, 1991.

[10]   Danish Recycling Cooperation. "Report on Visit to Ankara and Erzincan concerning the Erzincan Rehabilitation and Reconstruction Project". DEMEX Consulting Engineers, Denmark, 1992.

5

REUSE OF BUILDING MATERIALS AND DISPOSAL OF STRUCTURAL WASTE MATERIAL

Carlo De Pauw, Belgian Building Research Institute

5.1 Introduction

Over the last decades concern for the environment has steadily increased on both social and political levels. Gradually the economic world has also become concerned, mainly as a result of external pressure. The construction industry in particular has more and more problems in disposing of its construction and demolition waste because tipping charges and taxes for dumping the waste are increasing, due to the growing concern for the environment.

Many studies and conferences have already taken place, mainly in Europe, Japan and the United States, concerning the feasibility of the reuse of building waste. At present technical know-how is sufficient to implement the methods of reuse that have been studied. In some densely populated European countries, where dumping is becoming more and more difficult, the recycling of debris has already begun. A survey by the European Demolition Association (EDA) in 1992 [5], indicated that at present, the following numbers of recycling plants are in operation in the main European countries: 60 in Belgium, 20 in Denmark, 50 in France, 220 in Germany, 70 in the Netherlands, 43 in Italy and 120 in Great Britain.

However, no matter how strong the social concern about the environment may be, and disregarding the technical possibilities, the final motivation for recycling construction and demolition waste will always remain an economic one. Generally the economic interest in recycled building materials is governed by three factors:

- The availability, and thus the cost, of natural building materials.
- The availability of disposal space, the tipping charges and the taxes for dumping construction and demolition waste.
- The transportation costs.

Regional variations in these factors have of course a major impact on the willingness for recycling debris. This can be illustrated with two examples.

The first example describes the recycling situation in Belgium. Between Belgium's regions, Flanders and Wallonia quite a number of differences are observed [3] in relation to the use of recycled materials. In order to appreciate these differences it is necessary to realise that there is a distinct geographical difference between the regions (Fig.1).

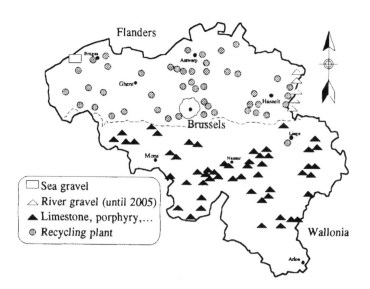

Fig. 1 : Map of Belgium [3]

Furthermore it should be realised that since the mid eighties the regions are autonomous to develop their own environmental and building policy and legislation.

Flanders, which is situated in the North of Belgium, has a very high population density and nearly no natural aggregate resources are available. The production of construction and demolition waste has been calculated at circa 4.6 million tons per year, i.e. 807 kg per year and per habitant. Tipping costs for building waste are typically in the range of 150 to 400 Belgian francs per ton. A high environmental tax of 350 Belgian francs per ton is added to this. At the end of 1992 new environmental legislation was introduced prohibiting the dumping of recyclable waste coming from construction and demolition. Taking this whole situation into account, it is quite logical that a well-established recycling industry is operating in Flanders and many new initiatives are currently going on.

Wallonia is situated in the South of Belgium and has a population density of only 190 habitants per square kilometre. A large number of quarries are dispersed over the territory. Estimates indicate a production of about 2 million tons of construction and demolition waste per year, i.e. 625 kg per year and per habitant. Tipping costs are lower than Flanders and typically in the range of 80 to 300 Belgian francs per ton. An environmental tax of 150 Belgian francs per ton is added to this. Although the number of authorised dump sites is very limited at the moment, only one recycling plant is operating. Obviously this leads to problems in the disposal of the waste. As a result of this situation a co-operation has been set up to establish an adequate network of dump sites for inert waste and the exploitation of stock sites for non-contaminated excavated soil (soil banks). Further, it aims at landscaping existing unauthorised dump sites and closed quarries. The promotion of recycling is also one of the goals.

It is obvious that the geographical differences between Flanders and Wallonia will result in a different economic interest in recycling between the regions.

A second example is a Danish feasibility study [4] on the recycling of building waste in Kuwait in 1990.

Kuwait has a production of 3 million tons of construction and demolition waste and expects at least an equally large annual quantity for the coming 10 years. This large pile of annual building waste causes a disposal problem for Kuwait. Until now this waste has been disposed of in landfill and, what is worse, disposed of by fly-tipping in the desert, along roads and in unsuitable gravel pits. The total amount of waste suitable for crushing and recycling was assessed to be close to 1.1 millions tons a year. However, in Kuwait it is not possible to finance a recycling project through taxes and tipping charges. Because of the enormous space available, the tipping charges at the landfills are minimal, and transportation costs are totally decisive in the choice of the disposal place. Therefore any recycling initiative in Kuwait must compete with similar products on the market. So it is necessary to aim at a high quality which results in high demands on the machinery and operation. Fortunately natural gravel aggregates of good quality are scarce and thus prices are high, making recycling feasible.

In post disaster conditions, however, environmental concern and the above mentioned economical factors lose their importance. Social and economic life is totally destabilized at once and National and Local authorities' main concern (after the relief phase) is the rehabilitation of the region. The destroyed region is left with an enormous amount of debris which has to be cleared away as soon as possible. Taxes and tipping charges, that were very decisive in normal conditions, are no longer considered. Finding suitable disposal space for the huge pile of debris causes enormous problems. This is certainly the case in densely populated regions and cities, as illustrated in 2.8 and 2.9.

Thus, recycling of the building waste after a disaster seems a feasible solution in the rehabilitation phase. On the one hand there is an enormous pile of building debris which is cleared away and generally transported to disposal sites. On the other hand there is an insatiable demand for new building materials necessary for the reconstruction of the area. Natural materials have to be transported to the reconstruction site. However, because the demand is high and the materials are not always available they sometimes have to be imported from far-away. Therefore, both the site-clearance and reconstruction involve high transportation costs, increasing the financial burden on the local authorities.

It is obvious that recycling can be a suitable solution for the enormous pile of debris caused by a disaster and that it can help the region in its post-disaster rehabilitation process. However very often the stricken region is situated in less developed countries with low technical know-how and financial backup. Therefore the international community should take its reponsibility. The set-up of recycling programs, which involve the sending of recycling plants, technical know-how and skilled engineers and technicians, should be integrated into global disaster relief. The organisation of such programs should be discussed on an international level.

5.2 Reuse of debris as part of the post-disaster rehabilitation

In the previous chapters, different phases of the post-disaster rehabilitation process that precede the reconstruction phase have been described. However, because of their importance for the development of a recycling project it is necessary to highlight some elements of these previous phases.

In Figure 2 the different areas to be considered in the recycling of rubble in a post disaster period and their connection are schematically represented.

In this chart the previous mentioned themes of assessment, demolition and site-clearance are immediately recognisable. In practice, in the turbulent period after a disaster, these activities are developed at almost the same time. Regarding the recycling of debris, the assessment and classification of the damage to buildings can give an idea (chapter 3) of the amount and the quality of the building waste that is to be expected. This information is necessary for planning and dimensioning the recycling plant(s), providing optimal equipment and choosing the best location to implement the plant(s). The assessment work, however, can be assisted by other means such as aerial photographs of the area. The latter, for example, was used for the assessment of the amount of debris after the earthquake in El Asnam [2].

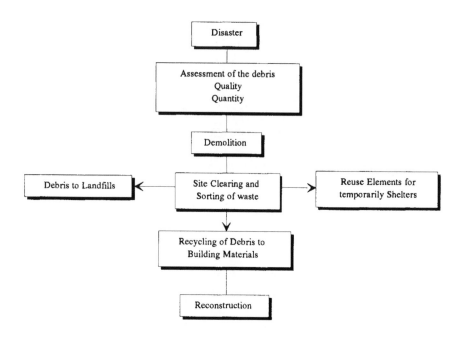

Fig. 2 : Post-disaster activities concerning recycling of debris

The quality of the debris is important as it affects the quality of the recycled products, and thereby the possibilities for reuse. Although the type of buildings indicate the quality of the debris, this quality will, in an important way, be determined by the demolition work (chapter 4). The most important factor is the presence of demolition experts and the availability of modern equipment.

During the recycling project in El Asnam [2] some hypotheses were advanced concerning the recyclability of the debris in relation to the degree of expertise of the demolition (Fig. 5).

Another important influence on the quality of the debris brought to the recycling plant is the method of site-clearance. Usually debris is collected quickly without sorting so rehabilitation can start inmediately. The building debris is therefore mixed with all kinds of other rubbish, which hampers the work at the recycling plant.

The construction and demolition waste can be mainly reused in three different manners:

- Immediate reuse of undamaged construction elements for building temporarily shelters. This depends on the availability of lifting equipment, as the rescue phase gets the highest priority in the use of them.
- Disposal of the debris to landfills, strengthening of river banks, or disposal as rubbish.
- Recycling of the debris as aggregates used for road construction or concrete building blocks.

Considering the use of recycled aggregates in structural concrete is not recommended because it is difficult, in post disaster conditions, to obtain the high quality of debris that is needed.

There are only very few examples of recycling projects in connection with larger site clearance and reconstruction projects following a damaging disaster. However, two interesting projects are reported below.

The first project is a pilot recycling operation that was developed by the Belgian Building Research Institute after a violent earthquake in El Asnam (Algeria) in 1980 [2]. The goal was to investigate the possibilities of recycling building debris into new concrete building blocks. The project illustrated the feasibility of recycling debris in situ. The description of this project gives a good idea of how to prepare a recycling project.

The second project concerns the installation of a recycling plant in Leninakan (Armenia) by the German Red Cross to help with the rehabilitation after the earthquake in 1988 [1]. It gives a good idea of the problems that accompany an intervention for recycling and reconstruction. The recycled aggregates were mainly used for road-works.

5.3 El Asnam, earthquake, Algeria, 1980

Introduction

The first example is a pilot project involving the recycling of building waste, undertaken by the Belgian Building Research Institute with the support of the Belgian government, after the earthquake in El Asnam on October 10th 1980. This extremely heavy earthquake and the many after shocks reduced the town and its surroundings to one big pile of rubble and caused thousands of casualties.

Based on the experience from the IRSIA-BBRI research (IRSIA = Institute for the encouragement of scientific research in industry and agriculture) which started in 1977 under the title "Demolition, recycling and dismantling of concrete", immediately after the disaster the BBRI proposed to set up a study on the recycling possibilities of the debris from El Asnam, to the Algerian authorities.

Not only was such an operation a large scale application of the research results, but it could also solve the problems of dumping on the one hand and the difficulty of finding sufficient quantities of building materials for the reconstruction of the town on the other.

The study was performed by the BBRI, with the financial help of the IRSIA and in close collaboration with a number of Algerian scientific institutes, control organizations and government agencies.

Damage caused by the earthquake

The main shock had a force of 7.3 on the Richter scale and was followed three hours later by an aftershock with a force of 6.0. After the two main shocks a very large number of relatively heavy aftershocks followed. These were responsible for aggravating the damage to the buildings.

After the earthquake the Algerian control authority made a thorough investigation of no less than 6,538 constructions in the devastated area of El Asnam. The investigation was performed by a hundred engineers over a two month period. The data was noted on specially developed forms.

To gain a picture of the total damage in El Asnam the assessment data were completed using aerial photographs of the area.

In view of the potential for recycling, the damage to buildings was obviously of most interest. Considering the seismic activity in the area it seemed prudent to repair only those buildings which sustained only minor damage, so that repairs could offer sufficient guarantee of further stability in the earthquake zone. In this respect the surveyed buildings could be subdivided into three groups as follows. A little over one third (35 %) of the buildings examined were simply recoverable with no or minor non-structural damage. Almost the same number of buildings (38 %) were destroyed or had to be demolished because of severe structural damage. The remainder (about 27 %) with minor damage to the whole of the building and only minor structural damage was considered reparable.

One of the remarkable features was that higher buildings were the most damaged. Of the buildings with 4 storeys (ground floor + 3) or more, more than 65 % had to be demolished.

Assessment of the building materials present

In order to gain a picture of the quantities of debris and of where this debris was located, one had, apart from the damage picture described before, to have data on the quantities and the categories of materials that were used in the different types of construction in El Asnam before the disaster.

There were three main types of buildings in the damaged areas, consisting of the materials presented in Figure 3.

It is interesting to consider that the city had been destroyed previously in 1954. About 23 % of the buildings in El Asnam were structures that survived that earthquake. So, almost three-quarters of the buildings were 25 years old or less.

The total quantity of materials used for El Asnam city was estimated at about 3 million tons spread over the different kind of materials (Fig. 4). The dwelling space alone represented approximately 1,440,000 tons of materials.

Local bricks (toub)	These bricks were used for very small houses and for part of the scattered farms.
Traditional masonry work	The masonry consisted of clay bricks, concrete blocks, local boulders, etc. It was used for houses with 1 or 2 storeys (G + 0, G + 1) or for small commercial buildings, made of reinforced or unreinforced loadbearing masonry work. This included about 75 % of the buildings in El Asnam.
Concrete	Mainly used as concrete skeleton with masonry infill work (bricks and concrete blocks) up to 5 storeys (G + 4), in dwellings (apartment buildings), individual dwellings (large and medium), office buildings, schools, hospitals, commercial buildings, etc. Furthermore, concrete was used for panels and slabs or for water towers, silos, construction works, etc.

Fig. 3 : Kinds of material used in the damaged areas

The apartment buildings within the group of dwellings represented a quantity of materials which was estimated at about 420,000 tons, including concrete (170,000 tons), concrete floor slabs (60,000 tons) and the infill masonry of bricks and concrete blocks (190,000 tons). The timber and other finishing materials were not included. These materials were located in about 450 apartment buildings spread over town in 23 groups of buildings, each consisting of 2 to 65 buildings per group, with a number of floors varying from 1 to 4. It represented approximately 1.4 tons materials for every m^2 floor surface of the apartment buildings.

The remainder of the group of dwellings consisted of houses or one family dwellings plus some small commercial buildings combined with a home, which represented approximately 1,020,000 tons of materials. This material consisted for the most part of masonry work (concrete blocks and bricks). Here also the timber and fittings in the buildings were not included. These materials were spread over 71 groups of houses, where a distinction was made between very large homes (mostly isolated homes with more than one floor and a total floor surface of approximately 320 m^2), large homes (approximately 240 m^2), medium large homes (approximately 80 m^2) and small homes (approximately 50 m^2), the latter two home types consisting of only a ground floor. These were mean and approximate figures. In evaluating the distribution of the quantities of materials over the town the building densities in the different areas were considered.

The quantity of materials in the remaining structures was roughly estimated at 1,500,000 tons. This covered administrative buildings, health care buildings, army barracks, hotels, schools and similar institutes, sports facilities and similar buildings, sociocultural buildings, mosques, industrial buildings, market buildings, railway stations, prisons, drinking water reservoirs, etc. Many buildings from this category roughly corresponded, from the point of view of the materials content, with the above-mentioned apartment buildings. Because of the diversity in the ground plan, number of floors, etc., only a rough estimation could be made for the approximately 750 structures spread over 50 groups.

| Building categories | Materials (tons) | |
	Concrete (slabs included)	Mixed materials (concrete, bricks and concrete blocks)
I - Apartment buildings	230,000	190,000
II - Dwellings	---	1,020,000
III - Collective and industrial buildings	800,000	700,000
Total	1,030,000	1,910,000
General total	2,940,000	

Fig. 4 : Quantities of materials used in El Asnam city

Considering a quantity of 1.4 tons of material per m² dwelling surface, this amounts to about 900,320 m² of dwelling surface. Considering also 9,3 m² dwelling surface per inhabitant, based on the mean figure of 65 m² floor surface per dwelling unit and an occupation of on average 7 persons per dwelling unit, this means for the estimated 900,320 m² of dwelling surface a total number of inhabitants for El Asnam city of 96,913. In reality there were approximately 110,000 inhabitants, making a difference of 13,000.

However, considering the inhabitants of the slums and the isolated farms, and the fact that apartment buildings and buildings partly used for housing were erroneously categorized in the administrative buildings, the estimated 1.4 tons of building materials per m² dwelling surface was acceptable for the aim put forward, i.e. to provide a rough estimation of the quantities of debris.

Demolition and debris evaluation

The combination of the data on the materials (quantities and location) with the data on the damage provided a picture of the quantities of debris and the locations where this debris would be available. However, the quality of the debris, and thus the usability for recycling, depended heavily on the quality of the local work force and the demolition equipment being used. Thus, in order to get a final picture of the total recycling operation for El Asnam, a number of considerations were necessary concerning the recyclability of the debris, depending on the demolition method used.

Considering the demolition to be executed by more or less expert operatives using efficient and modern demolition techniques, the following hypotheses concerning the recyclability of the debris were advanced (Fig. 5):

Building category	Recyclable materials per demolition type (in %)		
	A - Demolition by experts with adequate modern techniques	B - Demolition by experts with limited technical equipment	C - Demolition with inadequate techniques
I - Apartment buildings	80 %	60 %	40 %
II - Dwellings	30 %	10 %	0 %
III - Collective and industrial buildings	70 %	50 %	30 %

Fig. 5 : Recyclable materials per demolition type

Then, still within the framework of the hypothesis that structural damage must lead to demolition rather than repair and considering the real damage accounted for in the 100 different areas of El Asnam, it was possible to calculate the quantities of recyclable materials (Fig. 6). Consideration was again given to the hypotheses on the recyclability in relation to the degree of expertise of the demolition.

This meant that for the demolition type B (demolition by experts with limited technical equipment) about 700,000 tons of debris could be recycled, which equalled to more than 1 year of breaking work for 3 debris recycling installations working at 100 tons/hour each.

Building category	Recyclable materials per demolition type (in tons)		
	A - Demolition by experts with adequate modern techniques	B - Demolition by experts with limited technical equipment	C - Demolition with inadequate techniques
I - Apartment buildings	223,000	164,000	110,000
II - Dwellings	197,000	67,000	---
III - Collective and industrial buildings	652,000	465,000	281,000
Totals	1,072,000	696,000	391,000
% of the total quantity of materials in the buildings	37 %	24 %	13 %

Fig. 6 : Recyclable materials for El Asnam

Demolition

In total a little over 4,000 structures in El Asnam were identified for demolition. The demolition operation could be divided into three cases:

1. Fully collapsed buildings for which the operation could be limited to the demolition of some parts of the debris pile to form transportable elements

2. Semi collapsed or lopsided buildings for which the demolition work was very dangerous and required more expertise than the other categories. This included, for instance, the case where the ground floor of buildings had collapsed, so that the floors above were resting unstably on the ground floor. In this category also belonged the structures where the columns or beams were partly collapsed, without causing the collapse of the building as a whole.

3. Apparently intact buildings with a slightly damaged structure which were considered unrepairable because of the fact that they were located in earthquake sensitive areas.

The most suitable demolition techniques for El Asnam were, considering the above, the use of explosives, the thermal lance, the ball and/or the hydraulic or pneumatic apparatus on telescopic arms. However, for several reasons including the fact that many of the buildings were reoccupied in no time and that slums were erected in the devastated districts, the use of explosives was prohibited. The other possible techniques are summed up in Figure 7.

Demolition project	Advised demolition techniques
Ruins	- balling - oxygen cutter for fragmentation - bulldozer
Semi collapsed or unstable buildings	- balling - oxygen cutters - thermal lance - hydraulic arms - pneumatic apparatus on arms
Looking intact	- all current techniques

Fig. 7 : Possible demolition techniques

Irrespective of the technique that could be used, selective demolition was neither wished nor economical feasible, as it was decided to recycle the debris into building blocks. Structural recycled concrete was not considered because of the earthquake sensitive zone (see further).

Recycling of the debris into building blocks

In the United States of America, Japan and Western Europe research
had been performed for several years into the possibilities for reusing
building and demolition debris in new building materials. A large part of
this research was concentrated on the reuse of concrete debris as an
aggregate for new concrete, the so called recycled concrete. Here one
looked for techniques for economically justified separation of the
concrete from the reinforcement of reinforced or prestressed concrete
during the fragmentation.

From the results of this research it appeared that fragments from
concrete or masonry debris were suitable as large or even small
aggregates for new concrete, so long as a number of precautions were
taken with respect to the possible presence of impurities in the mother
concrete.

For lower applications such as foundation concrete, concrete blocks,
fence poles, binding concrete, etc., the total substitution of the natural
admixtures by recycled debris fragments was already feasible provided
that a number of precautions were taken.

For high quality applications such as the use of structural recycled
concrete the research results would no doubt be such that the green light
could be given on the technical level. A number of aspects concerning
the admissible contents of impurities, the precautions to be taken in the
stability calculations with recycled concrete, the economic feasibility,
the adaptation of the concrete constituents, the possible adaptation of the
rules of the mix design, the selective demolition for recycling purposes,
and the recycling installation itself, remained to be studied in the current
research.

Based on the qualitatively poor debris of El Asnam the production of
structural recycled concrete had to be excluded, both on economical
(separation techniques) and structural grounds (high requirements
concerning the earthquake resistance).

This, and the great shortage of building blocks, not only in El Asnam,
but also and in particular in the surrounding villages and rural areas,
were two important reasons for choosing the fabrication of building
blocks.

In addition, the recycling of the debris also seemed economically favourable. On the one hand there was a great need for natural aggregates for the reconstruction which had to be brought in from faraway places. Making the less demanding building blocks from recycled materials, and reserving the available natural aggregates for structural concrete, meant substantial savings on transport. On the other hand the enormous quantities of debris did not need to be dumped, which was an additional restriction on the transportation requirements.

However, a portion of the debris was unsuitable for recycling into building materials. In view of the great urgency of the first intervention immediately after the earthquakes part of the debris in the worst hit areas was put in piles and mixed with earth and lime in order to avoid epidemics. Part of the unsuitable debris (mixed debris) was used for the restoration and strengthening of the banks of the river Cheliff.
Nevertheless some hundreds of thousands of tons of debris were still available for breaking and sieving and subsequent use as aggregates for building blocks.

One of the conditions for recycling was the need for guarantees that with the average debris (i.e. without having to take too many precautions as to the separation of the different components) it would be possible to produce blocks of an acceptable quality, i.e. of about the quality of blocks made from natural aggregates, in local working conditions and using "normal" means (breaking facilities and block making machines). For these reasons tests were made using a debris mixture of bad quality in order to simulate the worst possible conditions (see further).

Pilot recycling operation

Materials used

By way of trial a number of concrete blocks or rather building blocks were produced from crushed debris coming from a group of 9 buildings (G + 3) which were constructed using a concrete skeleton with filling masonry consisting of concrete blocks and bricks with plastering (Figs 9 and 10).

This district was chosen because the demolition of these buildings was planned during the period of the investigation. Thus, after determining the quantities and qualities of the materials, demolition and recycling could be started right away.

The choice was furthermore influenced by the fact that these apartment buildings were fairly representative of a large number of the apartment buildings in El Asnam city, both in the design as well as in the materials used.

In general the concrete found in El Asnam was of poor to very poor quality. Apart from a doubtful compressive strength, there was honeycombing and other shortcomings. The mean compressive strength measured on 300 cores taken from different buildings after the disaster was 18 N/mm² with a standard deviation of 7 N/mm² [3].

The hollow concrete blocks used (length : 390 mm, width : 90,140 or 190 mm and height : 190 mm with a wall thickness of 25 to 30 mm) had the following composition: sand 0/2 (100 kg), gravel 3/8 (300 kg), cement P15-302 (50 kg) and a sufficient quantity of water to reach the consistency of "moist earth".

The bricks (length : 190 mm, width : 50,100 or 150 mm and height : 150 mm) were of the "Mediterranean Sea" type, with holes and had a mean compressive strength of 3.75 N/mm².

The external plastering was made up of mortar and was in most cases painted. The internal plastering was made up of plaster or mortar, in both cases with a thickness of about 10 mm.

The floor slabs (vaults or concrete slabs) were mostly covered with mortar.

The above mentioned materials were found in most of the city. For the chosen district as a whole, the materials quantities measured are presented in Figure 8.

Concrete debris	6,650 t
Concrete vaults and slabs	2,280 t
Mixed debris (concrete blocks + bricks+ plastering)	7,700 t
Total	16,530 t

Fig. 8 : The quantities of materials in the district

Fig. 9 : Partly collapsed building (G + 3)

Fig. 10 : Building debris used for recycling

Choice of the materials to be recycled

For a number of reasons the recycling operation was based on the worst possible debris in order to see whether the recycling was possible a priori, even without taking precautions during the removal of the debris. In view of the demolition methods used in El Asnam one could expect a total mixture containing all sorts of debris.

The aim was to use a large number of recycled blocks, for the local reconstruction in the surrounding villages in particular, where the quality requirements were lower.

For all these reasons the composition of Figure 11 was adopted for the production of the recycled concrete blocks.

Concrete debris	10 %
Concrete block debris	50 %
Brick debris	30 %
Impurities + miscellaneous debris (plastering, tiles, etc.)	10 %

Fig. 11 : Composition of the recycled concrete blocks (in percentage of weight)

Recycled building blocks

A first fragmentation of the debris was performed with sledge hammer and bulldozer. Afterwards the broken materials were fragmented to 0/25 mm in a secondary rotating crusher. The operation caused much dust because of the presence of brick masonry debris with plaster. Moreover many of the fragments resulting from the bricks came out of the crusher in the form of lamellae.

With the recycled aggregates thus obtained (Fig. 12), six series of building blocks (Fig. 13) were made, the compositions of which are shown in figure 14. In the first instance reference blocks were made based on natural aggregates and this, as for all other mixtures, under local conditions, i.e. heavy sunshine, with local production means and local workers. This procedure was chosen in order to approximate as near as possible the fairly bad characteristics due to the local factors, in order to reach the absolute minimum strength of the blocks.

Series nr.	Batching (kg) [1]				
	Cement P15-302	Natural Sand 0-2	Natural coarse aggregates 3-8	Debris	
				0-25	3-12
Reference	50	100	300	---	---
1	50	150	---	250	---
2	50	100	---	200	---
3	50	200	---	200	---
4	50	150	---	---	300
5	50	150	---	---	250
6	50	200	---	---	200
(1) Composition for the manufacturing of 24 hollow blocks of 390 mm x 190 mm x 190 mm					

Fig. 14 : Composition of the building blocks

The addition of the water as well as the vibrating of the block making machine were adapted to the consistency of the moment and to the customs of the local workers.

The drying occurred during sunshine and dry wind and regular water spraying was used as was typical in these areas.

The blocks were broken after 28 days in the LNTPB (Algerian public works laboratory) under the supervision of the engineers of the INERBA and the BBRI.

The blocks were leveled on top and bottom with quick hardening mortar. 12 reference blocks, 12 blocks of series 6 and 6 blocks of each of series 1 to 5 were tested. The results are given in Figure 15.

Fig. 12 : Recycled aggregates

Fig. 13 : Recycled building blocks

Reference blocks	Recycled blocks					
	series 1	series 2	series 3	series 4	series 5	series 6
3.6	2.6	2.6	3.3	3.4	4.6	1.9
3.9	3.3	2.6	3.2	3.8	6.4	1.8
3.9	1.9	4.0	2.7	4.5	7.1	2.3
3.7	1.6	4.4	3.3	4.8	9.2	1.9
4.5	1.7	3.0	3.0	3.1	8.7	2.9
2.9	2.2	2.9	2.0	4.9	6.1	2.4
3.9	---	---	---	---	---	2.0
2.1	---	---	---	---	---	1.4
4.6	---	---	---	---	---	2.3
2.9	---	---	---	---	---	2.3
4.0	---	---	---	---	---	2.3
3.7	---	---	---	---	---	1.4
3.7	2.2	3.3	2.9	4.1	7.0	2.1
Mean values						

Fig. 15 : Compressive strength (in N/mm²) of the gross surface of the blocks

Conclusions

The BBRI research on demolition and reuse of concrete and the pilot operation executed in El Asnam, have shown that such a recycling operation is possible. The results obtained were very convincing and proved that it was indeed feasible to recycle the debris as aggregates in concrete. However, although realistic the project was halted because for both political and mainly sociological reasons, i.e. opposition of the local people to the use of debris, under which some of their beloved had died, as a construction material.

Expert demolition, followed by a sufficient crushing capacity, could have resulted in over a million tons of recyclable material for the town of El Asnam, which roughly corresponds to some 50 million recycled building blocks. It would certainly have been an appropriate alternative to dumping several hundreds of thousands of tons of debris and the hauling over great distances of natural aggregates for the reconstruction of the town.

5.4 Leninakan, Armenia, 1988 [1]

This most recent example concerns the installation by the German Red Cross of a crushing plant in connection with the clearing and reconstruction work after the Armenian earthquake from December 1988. After this heavy earthquake, that destroyed more than 400 villages and caused 25,000 casualties and more than half a million homeless people, the German Red Cross ordered the installation of a recycling plant for debris, to support the suffering region of Leninakan.

The local conditions for installation were very difficult: 1600 metre altitude, severe winters, hot summers and heavy rainfall in spring and autumn, bad roads and transport conditions, insufficient water and electricity supply, etc. Another fact was that most of the debris - at the time of arrival of the recycling plant - was already transported and temporarily disposed of. This meant that the building debris was mixed with all kinds of waste, from household equipment to dead pets.

A particular problem was the transportation of the recycling plant to Leninakan. The equipment was transported by plane, with a Soviet Antonov An-124, to Eriwan. In January 1990, after being immobilised in Eriwan, everything was transported by trucks to Leninakan, 150 km away.

Without sufficient lifting machinery and with a lack of skilled local workers the German team managed the installation in seven weeks, and handed over to the Armenian authorities on the 15th of May 1990. One German stayed on stand-by for the maintenance of the equipment.

The plant was designed for a capacity of 120 t/h but achieved an output of more than 200 t/h and worked with no major problems. The produced construction material was primarily used for roadwork. A final report on this project is not yet published.

The whole installation covered a surface area of 13,700 m^2 and was able to resist new earthquakes because practically no foundation was provided. The whole recycling-activity covered more than 30,000 m^2.

The different successive working-phases that took place in the recycling process are as follows (Fig. 16):

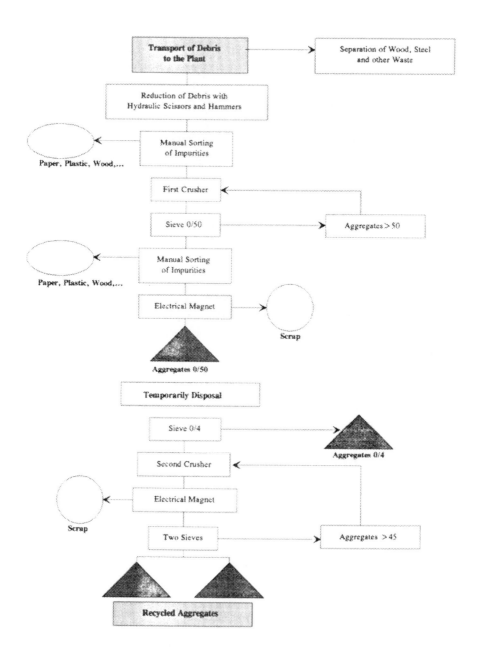

Fig. 16 : Recycling plant of Leninakan (Armenia)

- From the 150 tons of debris that was transported to the plant every hour, 30 tons of wood, steel and other waste was separated. Thus the crushing operation was fed with 120 tons of debris every hour.
- Hydraulic scissors and hammers and transportable scissors reduced the dimensions of the concrete and stone debris to suit them for the first crusher.
- During the transportation to this crusher a workman sorted the impurities.
- After crushing, aggregates with a diameter from 0 to 50 mm were separated by sieving. Next, another workman sorted the debris again, before it was passed under magnets to separate the iron. Then the debris was temporarily disposed of.
- After the temporary disposal the debris was separated from the finest aggregates with a diameter from 0 to 4 mm. The rest was crushed again in a secondary crusher.
- Then again the remaining iron was removed using magnets, before the remaining debris was separated in different fractions by two sieves. Granulates bigger than 45 mm were carried back to the secondary crusher.

5.5 Conclusions

Different studies over the last ten years have proved the technical feasibility of recycling debris into new building materials. Recycling projects have already been developed mainly in Japan, the United States and different European countries. The economic feasibility of these projects is mainly based on transportation costs, tipping fees and environmental taxes.

In regions that are destroyed by a disaster, recycling is highly suitable because enormous amounts of debris meet with an insatiable need for new construction materials. Recycling programs set up on an international level, providing the stricken region with the necessary equipment and know-how, could help the region in its rehabilitation process.

The feasibility of post-disaster intervention, considering the recycling of debris, is illustrated by the few examples (El Asnam and Leninakan) that have taken place.

However, these pilot projects have also illustrated that many other problems can occur related to local conditions. The main problems that occur for recycling projects after a disaster can be summarized as follows:

- Transportation and installation of the recycling plant(s) and other necessary equipment.
- Local conditions such as climate, infrastructure, building culture, etc.
- The absence or the lack of skilled local workmen.
- The urgency of the site clearance leads to the temporary disposal of the debris mixed with other waste.
- The covering of mixed debris with earth, lime, etc. to avoid epidemics. This makes the debris unsuitable for recycling.
- Social and cultural barriers for the acceptance of the idea of recycling disaster debris.
- Political barriers.

Valuable initiatives can be obstructed by important political, cultural and social barriers. In El Asnam such barriers were decisive for the failure of the project.

If we want to break through these barriers it is necessary to discuss the possibilities for the recycling of debris following a disaster on an international level, and to consider the setting up of an international organisation concerning this matter. On this international platform it would be easier to come to political agreements, to discuss cultural differences and to inform the population about the necessity and the feasibility of recycling debris. It would also make it easier to organise and finance new recycling projects in regions stricken by a disaster. The latter are necessary in order to be able to collect more information for the development of guidelines for recycling of construction and demolition waste following disasters.

In the wider scope of site clearance following a disaster, consideration on an international level is also necessary between the different authorities, organisations and professions that work in the different activities that are developed. A recycling project, that can be seen as the last link in the chain of activities, will only succeed if the previous activities, namely damage assessment and classification, demolition and site clearance, consider the possibilities of recycling debris and take the precautions that are necessary. A good co-operation and information interchange between the different operations will therefore be indispensable.

References

[1] Steinforth Hans, "Baustoff - Recyclinganlage in Leninakan", Baustoff Recycling + Deponietechnik, 2/91, April 1991.

[2] De Pauw Carlo, "Recyclage des Descombres d'une Ville Sinistree.", C.S.T.C. Revue N°4, December 1982, Belgian Building Research Institute, Brussels, Belgium.

[3] Vyncke J. and Rousseau E., "Recycling of Construction and Demolition Waste in Belgium, Actual Situation and Future Evolution", Belgian Building Research Institute, Brussels, Belgium, 1993.

[4] Lauritzen E. and Kristensen N., "Kuwait - Building Waste Handling, Recycling and Utilization", Demex Consulting Engineers & Ramboll & Hannemann Consulting Engineers and Planners, Denmark, August 1990.

[5] Buchner S. and Scholten L.J., "Demolition and Construction Debris Recycling in Europe'", European Demolition Association (EDA), 1992.

Appendix I

References on demolition and reuse of building materials in disaster relief

Dr. Ewald Andrews. Bevölkerungsschutspolitik im Nord-Süd-Dialog : IDNDR 1990 - 2000". Bundesverband für den selbsschutz, Postfach 20 01 61, 5300 Bonn 2, 1992.

Brundtland Commission Report. "Our Common Future". United Nations, 1987.

Carl Bartone, Janis Bernstein and Frederick Wright. "Investments in Solid Waste Management". Infrastructure and Urban Development Department, The World Bank, Washington, 1990.

Robert Bolin and Lois Stanford, "Shelter, Housing and Recovery: A Comparison of U.S. Disasters", DISASTERS Vol 15, No. 1, 1991, Basil Blackwell, Oxford, UK.

R. Bradley Burdick, Lawrence Livermore National Laboratory, "Summary of the September 19, 1985, Mexico earthquake", U.S. Department of Energy, USA, Jan 1986.

Buchner S. and Scholten L.J., "Demolition and Construction Debris Recycling in Europe", European Demolition Association (EDA), 1992

California Office of Emergency, "Heavy rescue, student manual", Federal Emergency Management Agency, Washington D.C., USA, 1980.

Michael A. Cassaro & Enrique Martinez Romero, "The Mexico Earthquakes-1985, Factors involved and lessons learned", American Society of Civil Engineering, New York, 1986.

CEB, "Diagnosis and Assessment of Concrete Structure", State of the Art Report of the Comité Euro-International du Béton, Information bulletin N° 192, Lausanne, January 1989.

CEB, "Assessment of Concrete Structures and Design Procedures for Upgrading (Redesign)", Comité Euro-International du Béton, Information bulletin N° 162, Lausanne, August 1983.

Centre for Advanced Engineering. "Lifelines in Earthquakes - Wellington Case Study". The University of Canterbury, New Zealand, 1991.

Colby, Michael E., Environmental management in development countries. World Bank discussion papers 80, 1990.

Dangroup International A/S, "Jordskælvssikring af bygningskonstruktioner i Algeriet", Det danske Boligministerium, København, DK, 1985.

Danish Recycling Cooperation. "Report on Visit to Ankara and Erzincan concerning the Erzincan Rehabilitation and Reconstruction Project". DEMEX Consulting Engineers, Denmark, 1992.

De Pauw Carlo, "Recyclage des Decombres d'une Ville Sinistrée", C.S.T.C. -Revue/N°4/ Dec 1982, Bruxelles, 1982.

Richard A. Dick, "S.E.E. Members Aid Mexico in its Time of Need", The Journal of Explosives Engineering Volume 3, Number 4, Nov./Dec. 1985, The Society of Explosives Engineers, Dublin, Ohio.

Disaster Prevention and Management, Vol. 1, Nos. 1 & 2, 1992, MCB University Press, Bradford UK.

David J. Dowrick, "Preliminary Field Observations of the Chilean Earthquake of 3 March 1985", Bulletin of the New Zealand National Society for Earthquake Engineering, Vol. 18, No 2, June 1982.

Frances D'Souza, "Recovery following the South Italian Earthquake, November 1980 : Two contrasting examples", DISASTERS Vol 6, No. 2, 1990, Foxcombe Publications, London, UK.

Claës Dyrbye, "Jordskælvsteknik" (Earthquake dimension), Technical University of Denmark, Copenhagen, Serie F, No. 125, 1991.

Economic Commission for Europe Committee on Housing, Building and Planning, "Second Seminar on Construction in Seismic Regions", Lisbon, Portugal, October 1981.

EEFIT Field Report, "The Luzon, Philippines Earthquake of 1990", EEFIT, Great Britain, June 1991.

Explosive Engineers, Inst. of, "REDR - Engineers for Disaster Relief", Oct. 1990.

"Field Manual: Postearthquake Safety Evaluation of Buildings", Applied Technology Council (ATC 20-1)

John Geipel, "Disaster and Reconstruction", George Allen & Unwin Ltd., UK 1982

Green, Norman B.,"Earthquake Resistant Building Design and Construction", Elsevier, New York, 1987.

J. Eugene Haas, National Academy of Sciences, "The Western Sicily earthquake of 1968", Colorado Univ., 1969.

Haas, Kates & Bowden, "Reconstruction Following Disaster", MIT Press, Massachusetts, 1977.

John Handmer & Dennis Parker, "British Disaster Planning and Management: An Initial Assessment", DISASTERS Vol 15, No. 4, 1991, Basil Blackwell, Oxford, UK.

Alan Hooper, "Dateline: Mexico City", The Journal of Explosives Engineering Volume 3, Number 4, Nov./Dec. 1985, The Society of Explosives Engineers, Dublin, Ohio.

Japan International Cooperation Agency, "Report of Japan Relief Team on Earthquake at Spitak, Armenia, USSR", February 1990.

Japan International Cooperation Agency, "Report of Japan Relief Team on Earthquake in Philippines of July 16, 1990", August 1990.

Kasai Y., "Report on the experiments for protection methods from flying objects in the demolition of urban structures by blasting". All Japan Association for Security of Explosives, 1990.

Kasai, Y. (Editor), "Demolition and Reuse of Concrete and Masonry". Proceedings of the Second International RILEM Symposium, Tokyo, Japan. Vol. I. Demolition Methods and Practice. Vol. II. Reuse of Demolition Waste. Chapman and Hall, London, UK 1988.

Kasai, Y., "Japanese Working Group for the RILEM 121-DRG, Task Force 2 Report", Nihon University, 1991.

Alcira Kreimer & Mohan Munasinghe, "Managing Natural Disasters and the Environment", Environment Department, World Bank, Washington, 1990.

Alcira Kreimer, "Rebuilding Housing in Emergency Recovery Projects", Policy and Research Division, World Bank, Washington, 1990.

Alcira Kreimer & Michele Zador, "Colloquium on Disasters, Sustainability and Development: A Look at the 1990's", Environment Department, World Bank, Washington, 1989.

Frederick Krimgold. "Pre-disaster planning". Department of Architecture, KTH Stockholm, Vol. 7, Sweden, 1974.

Richard Land, "The Morgan Hill Earthquake of April 24, 1984 - Summary of Highway Bridge Damage", Earthquake Spectra, Vol. 1, No. 3, 1985.

Erik K. Lauritzen & Martin B. Petersen. "Rehabilitation & Repair of Concrete -Partial Demolition by MINI-BLASTING". Concrete International, June 1991.

Erik K. Lauritzen, Niels K. Madsen & Jens Jensen. "Demolition of Motorway Bridge, Great Belt Link - A Research & Development Project on Fragmentation and Recycling of Reinforced Concrete 1991-92". DEMEX Consulting Engineers A/S. Denmark, 1992.

Lauritzen E. and Kristensen N. , "Kuwait-Building Waste Handling, Recycling and Utilization", Demex Consulting Engineers and Ramboll & Hannemann Consulting Engineers and Planners, Denmark, August 1990

Munich Reinsurance Company, "Earthquake Mexico '85", Münchener Rück, München, 1986.

Anthony Oliver-Smith, "Post-Disaster Housing Reconstruction and Social Inequality: A Challenge to Policy and Practice", DISASTERS Vol 14, No. 1, 1990, Basil Blackwell, Oxford, UK.

Anthony Oliver-Smith, "Successes and Failures in Post-Disaster Resettlement", DISASTERS Vol 15, No. 1, 1991, Basil Blackwell, Oxford, UK.

Organisation for Economic Co-operation and Development Report. "The State of the Environment". OECD, Paris, 1991.

S.V. Polyakov, "Design of Earthquake Resistant Structures", MIR Publishers, Moscow, 1983.

Antonios Pomonis, "The Spitak (Armenia, USSR) Earthquake: Residential Building Typology and Behaviour", DISASTERS Vol 14, No. 2, 1990, Basil Blackwell, Oxford, UK.

A. Pomonis, et al., "Assessing human casualties caused by building collapse in earthquakes", International Conference on the Impact of Natural Disasters, UCLA, USA, July 1991.

Frank Press, "The Role of Science and Engineering in Mitigating Natural Hazards", Bulletin of the New Zealand National Society for Earthquake Engineering, Vol. 18, No 2, June 1982.

Price, Lindsell & Buchner, "IABSE Colloquium Bergamo 1987 - Monitoring of a post-tensioned bridge during Demolition", IABSE, Zürich, 1987.

Proceedings of the Seventh European Conference on Earthquake Engineering, Technical Chamber of Greece, Athens, Greece, September 1982.

Robert K. Reitherman, "Earthquake, What to do - and why", California Geology, Volume 35, Number 3, California Division of Mines and Geology, March 1982.

Christoffer Rojahn, ATC, "Earthquake Damage Evaluation Data for California", Federal Emergency Management Agency, California 1985.

Christopher Rojahn, "Procedures for Postearthquake Safety Evaluation of Buildings", Applied Technology Council (ATC-20)

Schuppisser S. and Studer J., "Earthquake Relief in Less Industrialized Areas", A.A.Balkema; International symposium organized by the Swiss national committee for earthquake engineering and the Swiss society of engineers and architects, Zürich, Switzerland, March 1984.

Alan Scott, "Preparing for the BIG ONE in Central U.S.", EQE International Review, San Francisco, Spring 1991.

P.J. Smith, "Training for Crisis Management", DISASTERS Vol 14, No. 1, 1990, Basil Blackwell, Oxford, UK.

Hans Steinforth, "Baustoff-Recyclinganlage in Leninakan", BAUSTOFF RECYCLING + DEPONIETECHNIK 2/91, April 1991, STEIN-Verlag Baden-Baden GmbH, D.

Dipl.-Ing. Steinforth, "Einsatz einer Baustoff-Recycling-Anlage im Erdbebengebiet Armenien/UdSSR". 6. Symposium Recycling-Baustoffe, Cuxhaven, Germany, 1990.

H. Tiedemann, "Quantification of Factors Contributing to Earthquake Damage in Buildings", Engineering Geology, Elsevier, 1984.

Tova Maria Solo, "Dealing with Disasters", The Bank's World, published in Jan. and Feb. 1989 issues.

UPI, "Earthquake 7.1 San Francisco Bay Area October 17, 1989", LTA
Publishing Company, Portland, Oregon, 1989.

Urban Edge, "Mexico City: A remarkable Recovery", Vol. 11, No. 8, Oct.
1987.

Urban Edge, "Recovering from Sudden Natural Disasters", Vol. 10, No. 10,
Dec 1986.

Urban Edge, "Lessons in Disasters and Development", Vol. 8, No. 10, Dec.
1984.

Vyncke J. and Rousseau E., "Recycling of Construction and Demolition
Waste in Belgium, Actual Situation and Future Evolution", Belgian
Building Research Institute, Brussels, Belgium, 1993

Major D M Webb and Captain P D Cook, "Earthquake Relief in Mexico
City", The Royal Engineers Journal Volume 100, No. 4, December
1986, Chartham, Kent, UK.

Georg E. Wickham, Defense Civil Preparedness Agency, "What the planner
needs to know", NTIS, Washington, 1975.

World Bank Conference. "Environmental Management and Urban
Vulnerability". Environment Department, The World Bank,
Washington, 1992.

Richard N. Wright, National Bureau of Standards, "Building performance in
the 1972, Managua earthquake", Washington, D.C., 1973.

Peter I. Yanev, "Effects of the Miyagi-Ken-oki Earthquake on Reinforced
Concrete Structures", Concrete International, November 1981.

Appendix II

Summary of Key References

California Office of Emergency, "Heavy rescue, student manual",
Federal Emergency Management Agency, Washington D.C., USA,
1980.
A manual covering extensive information concerning search and rescue
operations in disaster situations ranging from earthquakes to snow-
slides, from high rise buildings to excavation equipment appropriate for
use in rescue operations.

Michael A. Cassaro & Enrique Martinez Romero, "The Mexico
Earthquakes-1985, Factors involved and lessons learned",
American Society of Civil Engineering, New York, 1986. Papers
presented at the International Conference held in Mexico City,
September 19-20, 1986.
Interesting paragraph: "**Local materials** - Local sources of good
aggregates for concrete were depleted many years prior to 1985.
Even the "good" aggregates of decades ago had smaller unit
weight and modulus of elasticity than those acceptable by world
standards; these features became more pronounced in recent years.
Unit weights over 10% below world standards and moduli of
elasticity as low as 57% of those standards were common. There
are doubts about the behaviour of such concrete in some modes of
failure when subjected to seismic loads. A research project is
being carried out to aid in interpreting earthquake effects on
existing structures and the Federal District Department has issued
provisions permitting only the better aggregates in structures. Such
aggregates can be brought in from nearby locations or obtained by

washing local materials.
Extract from "Emergency Regulations and the New Building Code" by
Emilio Rosenblueth, Hon. M. ASCE.

C. De Pauw, "Recyclage des Decombres d'une Ville Sinitree",
C.S.T.C. -Revue/N°4/ Dec 1982, Bruxelles, 1982.
This article describes work with crushing and recycling of concrete
from selected buildings after the El-Asnam earthquake, Algeria 1980.
The influence of pollution in the concrete and the poor quality concrete
affected the final result of produced concrete slabs, these being tested
for strength.

A description of the connection between the potential recyclable
amounts of building waste and the actual recycled amounts, resulting in
an increase of amounts for recycling with the implementation of more
advanced demolition methods and better educated manpower. Also the
benefits of recycling with reference to transport costs of raw materials
and aggregates.

Also a short report on political and cultural barriers for the
completion of the recycling of building waste project.

Richard A. Dick, "S.E.E. Members Aid Mexico in its Time of Need",
The Journal of Explosives Engineering Volume 3, Number 4,
Nov./Dec. 1985, The Society of Explosives Engineers, Dublin,
Ohio.
This publication reinforces the problems that may be experienced
concerning demolition techniques in some countries. Here the use of
explosives was not allowed for the rapid demolition of four selected
critically unstable buildings in Mexico City following the Mexico
Earthquake. This use of explosives would minimize the hazards to
labourers working on demolishing buildings by introducing more rapid
and safer working environments. The political decision was, however,
taken due to the fear that eventual survivors in the rubble could be
harmed and bodies could be further buried.

*Frances D'Souza, "Recovery following the South Italian Earthquake,
    November 1980: Two contrasting examples", DISASTERS Vol 6,
    No. 2, 1990, Foxcombe Publications, London, UK.*
Recovery and reconstruction in two communities following the South
Italian earthquake is described and differences are related to factors
such as pre-disaster development levels, relief decisions made during
the emergency phase, the kind of material aid received, local leadership
and economic opportunities. The combination of appropriate aid and
effective leadership appears to a potent force for recovery.

*EEFIT Field Report, "The Luzon, Philippines Earthquake of 1990",
    EEFIT, Great Britain, June 1991.*
Well illustrated and detailed report on the damage to constructions from
the Luzon Earthquake. EEFIT (Earthquake Engineering Field
Investigating Team) conducted a preliminary investigation two weeks
after the actual earthquake. The report covers the performance of
different building types and recommendations.

*"Field Manual: Postearthquake Safety Evaluation of Buildings",
    Applied Technology Council (ATC 20-1)*
This field manual is a result of the above mentioned publication,
"Procedures for Postearthquake Safety Evaluation of Buildings.

John Geipel, "Disaster and Reconstruction", Allen & Unwin Ltd., UK.
The results from and discussion on an in-depth questionnaire survey
carried out after the Friuli 1976 Earthquake. Social aspects,
resettlement, reconstruction and political aspects are discussed and
recommendations made for future disaster management from
governmental establishments. The effects of external intervention, be it
national or international, is also covered.

*Green, Norman B., "Earthquake Resistant Building Design and
    Construction", Elsevier, New York, 1987.*
A publication mostly on earthquake engineering, but with chapters on
observation and analysis of earthquake damage to buildings and
strengthening of these damaged buildings.

Haas, Kates & Bowden, "Reconstruction Following Disaster", MIT
    Press, Massachusetts, 1977.
The publication covers the planning and control of rescue, clearing and
rebuilding work after large natural catastrophies. Scenarios are set up
to which planners and the responsible authorities are to respond. The
authors require that important decisions are not to be postponed and
decisions concerning changes in the urban regions must be taken
according to the range of damages.

The authors describe the working phases after a natural catastrophe
by rescue work, renovation, rebuilding, urban construction and
development.

The publication concerns the renewal of construction codes after
earthquakes with insight to making urban areas safer in ensuing
catastrophies. The authors therefore require that the current
construction codes are implemented, as these codes can be difficult and
time consuming to alter. These codes must however be updated after
the rebuilding phase so as to maximise their benefit to the ensuing
catastrophe.

Alan Hooper, "Dateline: Mexico City", The Journal of Explosives
    Engineering Volume 3, Number 4, Nov./Dec. 1985, The Society of
    Explosives Engineers, Dublin, Ohio.
This article describes the help action immediately after the Mexico
earthquake of 1985 and cooperation with the authorities with reference
to classification grades for damages to buildings.

Also referred to is the authority's apprehension with regard to the
use of explosives in the rescue operations.

Japan International Cooperation Agency, "Report of Japan Relief Team
    on Earthquake in Philippines of July 16, 1990", August 1990.
A report JICA field trip to Philippines with extensive classification of
buildings, dams and bridges. Also recommendations for restoration,
reconstruction and further maintenance are put forward. Emergency
rehabilitation and construction of bridges is covered and the whole
report is thoroughly illustrate with photographs.

Alcira Kreimer, "Rebuilding Housing in Emergency Recovery Projects", Policy and Research Division, World Bank, Washington, 1990.

Alcira Kreimer & Mohan Munasinghe, "Managing Natural Disasters and the Environment", Environment Department, World Bank, Washington, 1990.

A volume on the papers presented at a World Bank colloquium referring to environmental and disaster management. Concerning actual demolition and recycling there is no mention, however interesting articles on reconstruction projects from disasters are included. Definitions of prevention and recovery are drawn and illustrated with case studies and the two-way relationship between environmental degradation and vulnerability to disaster is explored.

Anthony Oliver-Smith, "Post-Disaster Housing Reconstruction and Social Inequality: A Challenge to Policy and Practice", DISASTERS Vol 14, No. 1, 1990, Basil Blackwell, Oxford, UK.

This paper deals with the social dimensions of housing reconstruction after disaster in the context of the long term effects of reconstruction after the Yungay, Peru Earthquake-Avalanche 1970. The author contends that post-diaster housing reconstruction must avoid rebuilding structures which reflect, sustain and reproduce patterns of inequality and exploitation.

Anthony Oliver-Smith, "Successes and Failures in Post-Disaster Resettlement", DISASTERS Vol 15, No. 1, 1991, Basil Blackwell, Oxford, UK.

Site, layout, housing and popular input are presented as crucial issues in the determination of success or failure in post-disaster resettlement. Case material from Turkey, Iran and Peru is presented to illustrate how failure to attend to these issues produces unsuccessful resettlement villages. The article ends with a brief consideration of resistance to resettlement and alternative policies.

Antonios Pomonis, "The Spitak (Armenia, USSR) Earthquake:
Residential Building Typology and Behaviour", DISASTERS Vol
14, No. 2, 1990, Basil Blackwell, Oxford, UK.
A total description of buildings present in the area affected by the
earthquake, ranging from residential to commercial, from masonry
buildings to reinforced concrete buildings. The prefabricated reinforced
concrete framed buildings were all affected and "no building of this
type in Leninakan and Spitak will escape demolition".

A. Pomonis, et al., "Assessing human casualties caused by building
collapse in earthquakes", International Conference on the Impact
of Natural Disasters, UCLA, USA, July 1991.
A paper presented at above mentioned conference with data regarding
buildings damaged in various disasters. For example Caldrian
Earthquake, 24 Nov. 1976, Turkey where 84% of buildings were
destroyed.

Christoffer Rojahn, ATC, "Earthquake Damage Evaluation Data for
California", Federal Emergency Management Agency, California
1985.
A report involving extensive data of the effects from the California
earthquake, including a literature survey of earthquake damage to
various common types of buildings, bridges, tunnels, and earth dams.

Christopher Rojahn, "Procedures for Postearthquake Safety Evaluation
of Buildings", Applied Technology Council (ATC-20)
This publication contains the objectives of the Disaster Emergency
Services Committee of the Structural Engineers Association of
Northern California. Covered in the publication is the development and
documentation of qualitative procedures and guidelines for safety
evaluation of buildings damaged in earthquakes and to provide
appropriate training and field manuals and manuals describing this
methodology.

S. Schuppisser, "Earthquake Relief in Less Industrialized Areas",
  A.A.Balkema, Zürich, Swiss, March 1984.
The International Symposium on Earthquake Relief in Less
Industrialised Areas covered various aspects of interest including
papers on reconstruction and resettlement after earthquakes, damage
estimations for optimal relief operations, and emergency decisions on
safety of buildings damaged by earthquakes.

Hans Steinforth, "Baustoff-Recyclinganlage in Leninakan", BAUSTOFF
  RECYCLING + DEPONIETECHNIK 2/91, April 1991, STEIN-
  Verlag Baden-Baden GmbH, D.
The publication gives a description of the work in transportation and
setup of a crushing plant in Armenia after the May 1990 earthquake.
Lack of supplies and poor quality roads were some of the difficult
conditions which were encountered in the damaged areas.
Transportation of the 150 tons plant was carried out with a Soviet
Antonov An-124 transport plane.
  Also included in the publication is a description of the sorting of
building waste materials and crushing of up to 200 tons concrete per
hour.

Major D M Webb and Captain P D Cook, "Earthquake Relief in Mexico
  City"; The Royal Engineers Journal Volume 100, No. 4, December
  1986; Chartham, Kent, UK.
The authors suggest the planning of the relief work after a disaster
involving urban areas should be split into four phases:

    Phase 1   Life Saving
    Phase 2   Emergency assistance provided as:
              Services, eg. Medical, Engineers and Helicopters
              Supplies, eg. Food, Accommodation, Medical Supplies
              perhaps delivered by Hercules aircraft
    Phase 3   Rehabilitation
    Phase 4   Reconstruction

Georg E. Wickham, Defense Civil Preparedness Agency, "What the planner needs to know", NTIS, Washington, 1975.
This publication concerns building waste materials after catastrophies. A description of the amounts and types of waste to be found in collapsed buildings, estimated from the size of the existing buildings and their functions.

Includes a depiction of the types of materials to be found and how the building waste materials will include a variation from installations to automobile parts.

RILEM Reports

1   Soiling and Cleaning of Building Facades
2   Corrosion of Steel in Concrete
3   Fracture Mechanics of Concrete Structures - From Theory to Applications
4   Geomembranes - Identification and Performance Testing
5   Fracture Mechanics Test Methods for Concrete
6   Recycling of Demolished Concrete and Masonry
7   Fly Ash in Concrete - Properties and Performance
8   Creep in Timber Structures
9   Disaster Planning, Structural Assessment, Demolition and Recycling
10 Application of Admixtures in Concrete
11 Interfaces in Cementitious Composites

RILEM Proceedings

1   Adhesion between Polymers and Concrete.
2   From Materials Science to Construction Materials Engineering
3   Durability of Geotextiles
4   Demolition and Reuse of Concrete and Masonry
5   Admixtures for Concrete - Improvement of Properties
6   Analysis of Concrete Structures by Fracture Mechanics
7   Vegetable Plants and their Fibres as Building Materials
8   Mechanical Tests for Bituminous Mixes
9   Test Quality for Construction, Materials and Structures
10 Properties of Fresh Concrete
11 Testing during Concrete Construction
12 Testing of Metals for Structures
13 Fracture Processes in Concrete, Rock and Ceramics
14 Quality Control of Concrete Structures
15 High Performance Fiber Reinforced Cement Composites
16 Hydration and Setting of Cements
17 Fibre Reinforced Cement and Concrete
18 Interfaces in Cementitious Composites
19 Concrete in Hot Climates
20 Reflective Cracking in Pavements - State of the Art and Design Recommendations
21 Conservation of Stone and other Materials
22 Creep and Shrinkage of Concrete
23 Demolition and Reuse of Concrete and Masonry
24 Special Concretes - Workability and Mixing

RILEM Recommendations and Recommended Practice

RILEM Technical Recommendations for the Testing and Use of Construction Materials
Autoclaved Aerated Concrete - Properties, Testing and Design

RILEM, The International Union of Testing and Research Laboratories for Materials and Structures, is an international, non-governmental technical association whose vocation is to contribute to progress in the construction sciences, techniques and industries, essentially by means of the communication it fosters between research and practice. RILEM activity therefore aims at developing the knowledge of properties of materials and performance of structures, at defining the means for their assessment in laboratory and service conditions and at unifying measurement and testing methods used with this objective.

RILEM was founded in 1947, and has a membership of over 900 in some 80 countries. It forms an institutional framework for cooperation by experts to:

- optimise and harmonise test methods for measuring properties and performance of building and civil engineering materials and structures under laboratory and service environments;
- prepare technical recommendations for testing methods;
- prepare state-of-the-art reports to identify further research needs.

RILEM members include the leading building research and testing laboratories around the world, industrial research, manufacturing and contracting interests as well as a significant number of individual members, from industry and universities. RILEM's focus is on construction materials and their use in buildings and civil engineering structures, covering all phases of the building process from manufacture to use and recycling of materials.

RILEM meets these objectives though the work of its technical committees. Symposia, workshops and seminars are organised to facilitate the exchange of information and dissemination of knowledge. RILEM's primary output are technical recommendations. RILEM also publishes the journal *Materials and Structures* which provides a further avenue for reporting the work of its committees. Details are given below. Many other publications, in the form of reports, monographs, symposia and workshop proceedings, are produced.

Details of RILEM membership may be obtained from RILEM, École Normale Supérieure, Pavillon du Crous, 61, avenue du Pdt Wilson, 94235 Cachan Cedex, France.

RILEM Reports, Proceedings and other publications are listed below. Full details may be obtained from E & F N Spon, 2-6 Boundary Row, London SE1 8HN, Tel: (0)71-865 0066, Fax: (0)71-522 9623.

Materials and Structures

RILEM's journal, *Materials and Structures*, is published by E & F N Spon on behalf of RILEM. The journal was founded in 1968, and is a leading journal of record for current research in the properties and performance of building materials and structures, standardization of test methods, and the application of research results to the structural use of materials in building and civil engineering applications.

The papers are selected by an international Editorial Committee to conform with the highest research standards. As well as submitted papers from research and industry, the Journal publishes Reports and Recommendations prepared buy RILEM Technical Committees, together with news of other RILEM activities.

Materials and Structures is published ten times a year (ISSN 0025-5432) and sample copy requests and subscription enquiries should be sent to: E & F N Spon, 2-6 Boundary Row, London SE1 8HN, Tel: (0)71-865 0066, Fax: (0)71-522 9623; or Journals Promotion Department, Chapman & Hall Inc, One Penn Plaza, 41st Floor, New York, NY 10119, USA, Tel: (212) 564 1060, Fax: (212) 564 1505.

Recycling of Demolished Concrete and Masonry

Edited by **T C Hansen**, Professor of Building Materials, Technical University of Denmark

It is becoming increasingly difficult and expensive for demolition contractors to dispose of building waste and demolition rubble. For environmental reasons, public authorities are looking for ways of reusing these materials. The purpose of this book is to make the construction industry and public authorities aware of the technical possibilities for recycling of demolished concrete and masonry. It also shows how localized cutting and partial demolition of concrete structures can be carried out.

This new RILEM Report contains state-of-the-art reviews on three topics: recycling of demolished concrete, recycling of masonry rubble and localized cutting by blasting of concrete. It has been compiled by an international RILEM Committee and draws on research and practical experience worldwide.

Contents: **Part I: Recycled aggregates and recycled aggregate concrete: third state-of-the-art report 1945-1989.***Professor T C Hansen.* Introduction. First state-of-the-art report 1945-1977. Second and third state-of-the-art reports 1978-1989. Terminology. Original concrete. Production of recycled aggregate. Quality. Mechanical properties. Durability. Properties and mix design. Production. Use of crushed concrete fines for other purposes than production of new concrete. Products, codes, standards, and testing methods. Economic aspects of concrete recycling. Energy aspects of concrete recycling. Practical case histories. Recycling of fresh concrete wastes. Conclusions and recommendations. Acknowledgements. Literature references. Appendix A. **Part II: Recycling of masonry rubble.***Dr R R Schulz, Dr Ch F Hendricks.* Introduction. Historical survey. Prospects. Walling materials. Masonry rubble. Preparation. Properties of crushed masonry aggregate. Fresh concrete - composition and properties. Properties of hardened concrete. Durability of crushed masonry concrete. Applications. Ground powder from masonry rubble as binder. Standards, guidelines and instructions for production of crushed masonry aggregates and crushed masonry concrete. Economic aspects of the recycling of masonry rubble. Conclusions. References. **Part III: Blasting of concrete: localized cutting in and partial demolition of concrete structures.***C Molin, E K Lauritzen.* Foreword. Summary. Background and objectives. The principles of disintegration. Blasting techniques. The effect on nearby concrete. The effect on the environment. A brief comparison with other methods. Practical examples of blasting. Fragmentation of reinforced concrete. Conclusions. The need for research and development. References. Index.

"The book is to be recommended as an authoritative and comprehensive overview of progress in the field." - *Advances in Cement Research*
"will be of particular use to demolition and recycling contractors, and to concrete technologists and ready-mixed concrete producers" - *Concrete Abstracts*

RILEM Report 6

June 1992: 234x156: 316pp, 76 line drawings Hardback: 0-419-15820-0: £45.00

E & F N Spon
An imprint of Chapman & Hall